Dependence Models via Hierarchical Structures

Bringing together years of research into one useful resource, this text empowers the reader to creatively construct their own dependence models. Intended for senior undergraduate and postgraduate students, it takes a step-by-step look at the construction of specific dependence models, including exchangeable, Markov, moving average and, in general, spatio-temporal models. All constructions maintain a desired property of pre-specifying the marginal distribution and keeping it invariant. They do not separate the dependence from the marginals, and the mechanisms followed to induce dependence are so general that they can be applied to a very large class of parametric distributions. All the constructions are based on appropriate definitions of three building blocks: prior distribution, likelihood function and posterior distribution, in a Bayesian analysis context.

All results are illustrated with examples and graphical representations. Applications with data and code are interspersed throughout the book, covering fields including insurance and epidemiology.

LUIS E. NIETO-BARAJAS is Full Professor and Head of the Department of Statistics at the Instituto Tecnológico Autónomo de México (ITAM). He was previously President of the Mexican Statistical Association (2020–2021). For his thesis, he won the Savage Award (2001) and Francisco Aranda Ordaz Awards (2002–2004).

INSTITUTE OF MATHEMATICAL STATISTICS MONOGRAPHS

Editorial Board

Yingying Fan (University of Southern California)
Po-Ling Loh (University of Cambridge)
Ramon van Handel (Princeton University)
Rahul Mazumder (Massachusetts Institute of Technology)

ISBA Editorial Representative

Arnaud Doucet (University of Oxford)

IMS Monographs are concise research monographs of high quality on any branch of statistics or probability of sufficient interest to warrant publication as books. Some concern relatively traditional topics in need of up-to-date assessment. Others are on emerging themes. In all cases the objective is to provide a balanced view of the field.

In collaboration with the International Society for Bayesian Analysis (ISBA), selected volumes in the IMS Monographs series carry the 'with ISBA' designation at the recommendation of the ISBA editorial representative.

Other Books in the Series (*with ISBA)

1. *Large-Scale Inference*, by Bradley Efron
2. *Nonparametric Inference on Manifolds*, by Abhishek Bhattacharya and Rabi Bhattacharya
3. *The Skew-Normal and Related Families*, by Adelchi Azzalini
4. *Case-Control Studies*, by Ruth H. Keogh and D. R. Cox
5. *Computer Age Statistical Inference*, by Bradley Efron and Trevor Hastie
6. *Computer Age Statistical Inference (Student Edition)*, by Bradley Efron and Trevor Hastie
7. *Stable Lévy Processes via Lamperti-Type Representations*, by Andreas E. Kyprianou and Juan Carlos Pardo
8. *The Conway–Maxwell–Poisson Distribution*, by Kimberly F. Sellers
9. *Brownian Motion, the Fredholm Determinant, and Time Series Analysis*, by Katsuto Tanaka
10. * *Dependence Models via Hierarchical Structures*, by Luis E. Nieto-Barajas

Dependence Models via Hierarchical Structures

LUIS E. NIETO-BARAJAS
Instituto Tecnológico Autónomo de México (ITAM)

Shaftesbury Road, Cambridge CB2 8EA, United Kingdom

One Liberty Plaza, 20th Floor, New York, NY 10006, USA

477 Williamstown Road, Port Melbourne, VIC 3207, Australia

314–321, 3rd Floor, Plot 3, Splendor Forum, Jasola District Centre, New Delhi – 110025, India

103 Penang Road, #05-06/07, Visioncrest Commercial, Singapore 238467

Cambridge University Press is part of Cambridge University Press & Assessment, a department of the University of Cambridge.

We share the University's mission to contribute to society through the pursuit of education, learning and research at the highest international levels of excellence.

www.cambridge.org
Information on this title: www.cambridge.org/9781009584111
DOI: 10.1017/9781009584128

© Luis E. Nieto-Barajas 2025

This publication is in copyright. Subject to statutory exception and to the provisions of relevant collective licensing agreements, no reproduction of any part may take place without the written permission of Cambridge University Press & Assessment.

When citing this work, please include a reference to the DOI 10.1017/9781009584128.

First published 2025

A catalogue record for this publication is available from the British Library

A Cataloging-in-Publication data record for this book is available from the Library of Congress

ISBN 978-1-009-58411-1 Hardback

Cambridge University Press & Assessment has no responsibility for the persistence or accuracy of URLs for external or third-party internet websites referred to in this publication and does not guarantee that any content on such websites is, or will remain, accurate or appropriate.

To my family: Lyn, Rodrigo, Diego and Santiago

QUIQUE

Contents

Preface		*page* ix
Acknowledgements		xiii
1	**Introduction**	1
1.1	Statistical Inference	1
1.2	Common Probability Distributions	4
1.3	Moments	11
1.4	Stochastic Processes	17
2	**Conjugate Models**	23
2.1	Definitions	23
2.2	Detailed Examples	24
2.3	Summary of Conjugate Models	27
3	**Exchangeable Sequences**	33
3.1	Definitions	33
3.2	Pre-specified Invariant Distributions	36
3.3	Generalisations	41
3.4	Applications	43
4	**Markov Sequences**	51
4.1	Definitions	51
4.2	Examples	54
4.3	Applications	63
5	**General Dependent Sequences**	70
5.1	First Attempts	70
5.2	Main Results	74
5.3	Particular Exponential Family Cases	79

6	**Temporal Dependent Sequences**	84
6.1	Moving Average Type of Order q	84
6.2	Markov versus $MA(q)$ Type	89
6.3	Seasonal Models	94
6.4	Periodic Models	100
7	**Spatial Dependent Sequences**	102
7.1	Latent Areas Model	102
7.2	Latent Edges Model	106
7.3	Model Comparison	109
7.4	Spatio-Temporal Models	113
8	**Multivariate Dependent Sequences**	118
8.1	Multivariate Models	118
8.2	Dirichlet Process Models	122
Appendix	**Data Sets**	125
References		132
Index		135

Preface

Motivation

I first encountered statistics when I was doing my undergraduate studies in actuarial science. After being frustrated because I did not get a job as tutor in statistics in my senior year, I got a part-time job in marketing research where I could really appreciate the importance of statistics in practice. This motivated me to follow the path of light into statistics.

While pursuing my PhD in statistics at the University of Bath, I basically developed two lines of research. These are construction of continuous and discrete-time stochastic processes to define prior distributions in the area of Bayesian non-parametric statistics.

During my second sabbatical leave in the Department of Statistics at Oxford University, in the academic year 2014–15, I was invited to give a graduate lecture. This consisted of a long seminar for research students, but nobody told me that (distinguished) professors were able to attend. So, I had the privilege of having Professor David Cox as a student in my class, who, fortunately for me, did not ask any questions.

The topic I chose for the lecture was the collection of contributions I had made in the development of discrete-time stochastic processes. Some students personally approached me and others sent me emails congratulating me on the lecture.

During my third sabbatical in the Department of Statistical Sciences at the University of Toronto, in the academic year 2023–24, I was also invited to deliver a series of research seminars for the postgraduate students. I decided to update my previous notes with my most recent findings, and this is when this book was born.

Although this book is mainly intended for (post) graduate students, undergraduate students who have taken courses in probability and stochastic processes may also be able to follow the ideas presented in the book.

The book is self-contained and required background is reviewed in the first chapters.

This book contains the results of more than 20 of my research papers. It explains step by step how to construct dependence models, starting from exchangeable, Markov, moving average and, in general, spatio-temporal stochastic processes. All constructions maintain a desired property of pre-specifying the marginal distribution and keeping it invariant, regardless of the type of dependence, which makes these series of constructions appealing.

A pre-specified marginal distribution in dependence models can be defined in different ways. One of them is through the use of copulas. These separate the dependence from the marginal distributions to define a joint model. On the other hand, the constructions presented here do not separate the dependence from the marginals, but the mechanisms we follow to induce dependence are general ones that can be applied to a very large class of parametric distributions. All constructions presented here are based on appropriate definitions of three building blocks: prior distribution, likelihood function and posterior distribution, in a Bayesian analysis context.

Contents Description

Brief descriptions of the contents of each chapter follow.

In Chapter 1 we start by reviewing the different types of inference procedure. Then we introduce notation by providing a list of the probability distributions that will be used later on, review some results on conditional moments and finish by introducing the most common discrete-time stochastic processes that show dependence in time and space.

In Chapter 2 we define what a conjugate family is in a Bayesian analysis context and develop detailed examples of some cases. We finish by providing a list of conjugate models.

In Chapter 3 we introduce the concept of exchangeability and show how to construct exchangeable sequences. We then present our first explanation of how to construct exchangeable sequences while maintaining a desirable marginal distribution and provide detailed examples. We finish with an application of exchangeable constructions in a metanalysis.

In Chapter 4 we describe a general procedure to construct Markov sequences with invariant distributions. We derive several examples in detail and finish with some applications in survival analysis.

In Chapter 5 we start with some attempts to construct dependence sequences with order larger than one and present a general method to achieve an invariant distribution via a three-level hierarchical model. We finally present some results involving exponential families.

In Chapter 6 we show how to define temporal dependent sequences using a moving average type of construction. We compare the performance of this construction with a Markov process type. We finally extend the models to include seasonal and periodic dependencies.

In Chapter 7 we present two spatial dependent models: one based on defining a latent variable for each area, and the other based on defining one latent variable for each pair of areas. We compare both models with a real data set. Extensions to spatio-temporal constructions are also considered.

In Chapter 8 we conclude the book by presenting dependent models for random vectors and for stochastic processes.

I hope you will enjoy the book as much as I enjoyed the construction of the models.

Literature Review

There is a series of books that deal with dependent models based on discrete- and continuous-time stochastic processes. These alternatives are useful for the analysis of real data sets.

The first topic to suggest is copula models. These models separate the dependence from the marginal behaviour of variables, starting from the classical book Nelsen (2006), to recent books like Joe (2023).

To understand the basics of the probability models used in this book, we suggest that the reader review the topic of discrete-time stochastic processes. For example, Collet (2018) contains general applications, and Gallager (2012) has an engineering and operational research context.

The first dependent model proposed here is an exchangeable model. This is also known as the general concept of hierarchical models. Books that discuss this topic are typically within the Bayesian paradigm; for instance, Gelman and Hill (2006) contains applications in social sciences, and Congdon (2019) has applications in R.

Other types of dependence models that are treated here are useful for time series analysis. The classic book by Chatfield and Xing (2019) studies time series models under a classical approach with applications in R, and Broemeling (2019) presents time series analysis following a Bayesian approach. A particular type of time series modelling is based on filtering and smoothing. Särkkä and Svensson (2023) deal with this topic and provide applications in Python and MATLAB.

Chapter 7 of this book presents models for spatial and spatio-temporal analyses. Diggle (2023) presents a classical approach with applications in R, and Haining and Li (2021) deal with spatial models under a Bayesian approach.

Acknowledgements

I would like to thank all my collaborators who have shared the road with me, starting with my BSc supervisor Mario Cortina, my MSc supervisor Eduardo Gutiérrez and my PhD supervisor Stephen Walker; my Mexican colleagues Enrique de Alba, Ernesto Barrios, Christian Carmona, Alberto Contreras, Ricardo Hoyos, Gabriel Huerta, Sergio Juárez, Ramses Mena, Manuel Mendoza, Gabriel Núñez, Carlos Pérez and Carlos Rodríguez; my Brazilian colleague Rodrigo Targino; my Chilean colleagues Alejandro Jara and Fernando Quintana; my Italian colleagues Antonio Canale, Riccardo Corradin, Antonio Lijoi, Bernardo Nipoti and Igor Prünster; my University of Texas MDACC colleagues Veera Baladandayuthapani, Neby Bekele, Yuan Ji, Peter Müller and Guosheng Yin; my University of Oxford colleagues Sarah Filippi, Chris Holmes and James Watson; my University of Toronto colleagues Radu Craiu, Ruyi Pan and Piotr Zwiernik. And all my students at ITAM.

1

Introduction

1.1 Statistical Inference

Probability is the way we quantify uncertainty. It is based on three axioms that develop the whole probability theory. We suggest the book Mood et al. (1974) to review the main concepts.

Statistics is the science of data. It is the science of collecting, exploring, presenting and making decisions from data. As a science, it is divided into two branches: descriptive statistics and inferential statistics. The former involves sampling and exploration, whereas the latter deals with decision making, which includes estimation, hypothesis testing and predictions.

To understand the previous definitions and some other concepts involved in statistics, let X denote a characteristic of interest that is measurable for individuals in a particular population. For instance, X could be the income, height or age of a person in a particular population (school, county, country, etc.). The *statistical population* is the collection of all possible values x_i for the individuals $i = 1, 2, \ldots$ of the population. In notation, $Pop = \{x_1, x_2, \ldots\}$, where the population size could be infinite. If we were able to obtain all possible values for the whole population, we could summarise them in a relative frequency table and plot it in a histogram, like the one depicted in Figure 1.1. If we make the histogram bins narrower and take the limit as the length of the bins goes to zero, by appropriately dividing by the bins' length we obtain a smooth curve such as the one plotted on top of the histogram in Figure 1.1. Let us denote this curve mathematically as $f(x \mid \theta_0)$, which is a function of x, the possible values of the variable of interest, and θ_0, which is the true population parameter. In other words, the population is fully characterised by the curve, that is, $Pop \iff f(x \mid \theta_0)$. The curve is actually a probability model or a density function for a random variable X indexed by the parameter θ_0.

In statistics, it is not common to have access to all population values, but usually we have access to a subset of them that we call a *sample*.

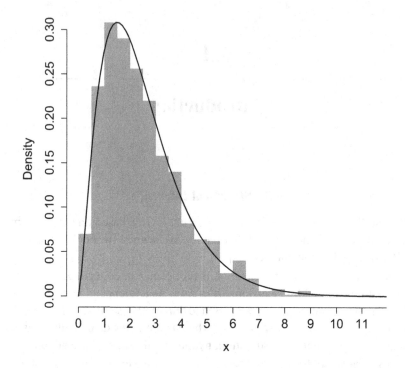

Figure 1.1 Histogram (shaded) and probability curve (solid line) for simulated data.

Formally, a sample is a finite collection of size $n < \infty$ of the characteristic of interest $\{X_1, X_2, \ldots, X_n\}$. Since these values are unknown to the researcher beforehand, they can be assumed to be (conditionally) independent random variables whose possible values are determined by the probability model $f(x \mid \theta)$, where the population parameter is usually unknown but belongs to a specific parameter space Θ, that is, $\theta_0, \theta \in \Theta$. For instance, for the data depicted in Figure 1.1, a possibility would be to assume a gamma model, namely $X \sim \text{Ga}(\alpha, \beta)$, where $\theta = (\alpha, \beta) \in \Theta = (\mathbb{R}^+)^2$.

There are two main approaches for statistical inference: classical or frequentist, and Bayesian. Within either inferential approach we could assume two possibilities for the population: a parametric assumption like the one we mentioned earlier where $X \sim f(x \mid \theta)$ and $\theta \in \Theta$; or a non-parametric assumption where the population is not characterised by a parametric model, namely $X \sim f(x)$ with $f \in \mathcal{F}$ and $\mathcal{F}$ the space of all probability models. This leads to four types of inferential procedures, which are summarised in Table 1.1.

1.1 Statistical Inference

Assumption\Procedure	Frequentist	Bayesian
Parametric	(1)	(3)
Non-parametric	(2)	(4)

Table 1.1 *Types of inferential procedures.*

We now briefly describe the generalities of each inferential procedure with a little more emphasis on (3) since most of the ideas discussed in this book belong to that context.

(1) **Frequentist Parametric**: The assumptions here are that observed data $\mathbf{X} = \{X_1, \ldots, X_n\}$ are a sample from population $X_i \sim f(x \mid \theta)$ of independent random variables where $\theta \in \Theta$. Sample information about θ is summarised in the joint distribution function $f(\mathbf{x} \mid \theta)$, which in the case of independent data is given by $\prod_{i=1}^{n} f(x_i \mid \theta)$, where, if seen as a function of θ, it is called likelihood. The frequentist inferential procedure is based entirely on likelihood, usually through maximisation. See, for example, Mood et al. (1974).

(2) **Frequentist Nonparametric**: The assumptions here are that observed data $\mathbf{X} = \{X_1, \ldots, X_n\}$ are a sample from population $X_i \sim f(x)$ of independent random variables and $f \in \mathcal{F}$. Sample information about f, or F, the corresponding cumulative distribution function (CDF), is summarised as $f(\mathbf{x}) = \prod_{i=1}^{n} f(x_i)$, which is the likelihood for f (and F). For instance, the maximum likelihood estimator (MLE) of F is the empirical distribution function $\widehat{F}(x) = \frac{1}{n} \sum_{i=1}^{n} I_{(-\infty, x]}(X_i)$, where $I_A(x)$ denotes the indicator function of set A that takes the value of one if $x \in A$ and zero otherwise. See, for example, Conover (1999).

(3) **Bayesian Parametric**: The assumptions here are that observed data $\mathbf{X} = \{X_1, \ldots, X_n\}$ are a sample from population $X_i \mid \theta \sim f(x \mid \theta)$ of conditional independent random variables and $\theta \in \Theta$. The word *conditional* is included because the Bayesian inferential procedure depends on an axiomatic theory that establishes that all unknown quantities must be quantified using the researcher's prior (uncertain) knowledge through $f(\theta)$. This prior knowledge is updated with the observed data through Bayes's theorem which states that

$$f(\theta \mid \mathbf{x}) = \frac{f(\mathbf{x} \mid \theta) f(\theta)}{f(\mathbf{x})}, \qquad (1.1)$$

where $f(\mathbf{x} \mid \theta)$ is the likelihood for θ and $f(\mathbf{x}) = \int_{\Theta} f(\mathbf{x} \mid \theta) f(\theta)$ or $f(\mathbf{x}) = \sum_{\theta \in \Theta} f(\mathbf{x} \mid \theta) f(\theta)$ is a normalising constant. Bayes's theorem is therefore a learning rule and $f(\theta \mid \mathbf{x})$ is called the *posterior distribution* for θ that contains all available information. To make decisions, the axiomatic theory establishes that preferences on consequences must be quantified by a utility (or loss) function which must be maximised (or minimised) after marginalising all uncertain quantities using the prior or posterior distribution, whichever is available. For instance, if we want to estimate θ with $\widehat{\theta}$ we could represent our preferences via a quadratic loss function $v(\widehat{\theta}, \theta) = a(\widehat{\theta} - \theta)^2$ for $a > 0$. If the posterior distribution is available, we obtain the expected loss as $\bar{v}(\widehat{\theta}) = \mathrm{E}\{v(\widehat{\theta}, \theta)\} = \int_\Theta a(\widehat{\theta} - \theta)^2 f(\theta \mid \mathbf{x}) d\theta$, from which, after minimisation, we obtain $\widehat{\theta} = \mathrm{E}\{\theta \mid \mathbf{x}\}$; that is, our point estimate for θ is the posterior mean. See, for example, Bernardo and Smith (2000).

(4) **Bayesian Nonparametric**: The assumptions here are that observed data $\mathbf{X} = \{X_1, \ldots, X_n\}$ are a sample from population $X_i \mid \theta \sim f(x)$ of conditional independent random variables and $f \in \mathcal{F}$. The axiomatic theory establishes that the researcher must quantify prior knowledge on f or F via $\mathcal{P}(f)$ or $\mathcal{P}(F)$. This is usually done via stochastic processes whose paths are densities or distribution functions. The two most typical choices are the Dirichlet process with precision parameter c and centring measure F_0, denoted by $\mathcal{DP}(c, F_0)$, see Ferguson (1973); and the Pólya tree with precision parameter c, variance function ϱ and centring measure F_0, denoted as $\mathrm{PT}(c, \varrho, F_0)$; see for example, Nieto-Barajas and Núñez-Antonio (2021). This prior distribution is updated with the observed data through Bayes's theorem (1.1), but adapted to stochastic processes, to obtain the posterior law $F \mid \mathbf{x}$. If we further represent our preferences via a quadratic loss function, the posterior point estimate for F will be $\mathrm{E}(F \mid \mathbf{x})$, which is known as the posterior predictive function. See Hjort et al. (2010).

Therefore, statistical procedures can be summarised as shown in the diagram of Figure 1.2. The arrow pointing down corresponds to descriptive statistics, whereas the arrow pointing up corresponds to inferential statistics.

1.2 Common Probability Distributions

In the following chapters we will use several common probability distributions as well as their first two moments. We summarise them here.

Discrete Distributions

- *Bernoulli distribution*: this is characterised by the following density:

$$f(x \mid \theta) = \theta^x (1-\theta)^{1-x} I_{\{0,1\}}(x),$$

1.2 Common Probability Distributions

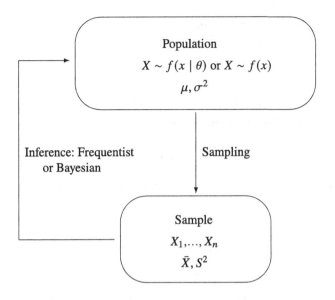

Figure 1.2 Diagram of statistics.

valid for $\theta \in (0, 1)$. This is denoted as Ber(θ). The first two moments are

$$E(X \mid \theta) = \theta \quad \text{and} \quad Var(X \mid \theta) = \theta(1 - \theta).$$

- *Binomial distribution*: this is characterised by the following density:

$$f(x \mid \theta) = \binom{n}{x} \pi^x (1 - \pi)^{n-x} I_{\{0,1,\ldots,n\}}(x),$$

with $\theta = (n, \pi)$ and valid for $\pi \in (0, 1)$ and $n \in \mathbb{N}$. This is denoted as Bin(n, π). The first two moments are

$$E(X \mid \theta) = n\pi \quad \text{and} \quad Var(X \mid \theta) = n\pi(1 - \pi).$$

- *Geometric distribution*: this is characterised by the following density:

$$f(x \mid \theta) = \theta(1 - \theta)^x I_{\{0,1,\ldots\}}(x),$$

valid for $\theta \in (0, 1)$. This is denoted as Geo(θ). The first two moments are

$$E(X \mid \theta) = \frac{(1 - \theta)}{\theta} \quad \text{and} \quad Var(X \mid \theta) = \frac{(1 - \theta)}{\theta^2}.$$

- *Negative Binomial distribution*: this is characterised by the following density:

$$f(x \mid \theta) = \binom{r+x-1}{x} \pi^r (1-\pi)^x I_{\{0,1,\ldots\}}(x),$$

with $\theta = (r, \pi)$ and valid for $\pi \in (0,1)$ and $r \in \mathbb{N}$. This is denoted as NB(r, π). The first two moments are

$$E(X \mid \theta) = \frac{r(1-\pi)}{\pi} \quad \text{and} \quad \text{Var}(X \mid \theta) = \frac{r(1-\pi)}{\pi^2}.$$

- *Poisson distribution*: this is characterised by the following density:

$$f(x \mid \theta) = e^{-\theta} \frac{\theta^x}{x!} I_{\{0,1,\ldots\}}(x),$$

valid for $\theta > 0$. This is denoted as Po(θ). The first two moments are

$$E(X \mid \theta) = \theta \quad \text{and} \quad \text{Var}(X \mid \theta) = \theta.$$

- *Beta-Binomial distribution*: this is characterised by the following density:

$$f(x \mid \theta) = \binom{n}{x} \frac{\Gamma(a+b)}{\Gamma(a)\Gamma(b)} \frac{\Gamma(a+x)\Gamma(b+n-x)}{\Gamma(a+b+n)} I_{\{0,\ldots,n\}}(x),$$

where $\Gamma(\cdot)$ is the gamma function that satisfies $\Gamma(a) = (a-1)\Gamma(a-1)$, with $\theta = (a, b, n)$ and valid for $a, b > 0$ and $n \in \mathbb{N}$. This is denoted as BBin(a, b, n). The first two moments are

$$E(X \mid \theta) = \frac{na}{a+b} \quad \text{and} \quad \text{Var}(X \mid \theta) = \frac{nab(a+b+n)}{(a+b)^2(a+b+1)}.$$

- *Beta-Negative Binomial distribution*: this is characterised by the following density:

$$f(x \mid \theta) = \binom{r+x-1}{x} \frac{\Gamma(a+b)}{\Gamma(a)\Gamma(b)} \frac{\Gamma(a+r)\Gamma(b+x)}{\Gamma(a+b+r+x)} I_{\{0,1,\ldots\}}(x),$$

with $\theta = (a, b, r)$ and valid for $a, b > 0$ and $r \in \mathbb{N}$. This is denoted as BNB(a, b, r). The first two moments are

$$E(X \mid \theta) = \frac{rb}{a-1}$$

if $a > 1$, and

$$\text{Var}(X \mid \theta) = \frac{rb(a+r-1)(a+b-1)}{(a-1)^2(a-2)}$$

if $a > 2$.

1.2 Common Probability Distributions

- *Gamma-Poisson distribution*: this is characterised by the following density:

$$f(x \mid \theta) = \frac{b^a}{\Gamma(a)} \frac{\Gamma(a+x)c^x}{x!(b+c)^{a+x}} I_{\{0,1,\dots\}}(x),$$

with $\theta = (a,b,c)$ and valid for $a,b,c > 0$. This is denoted as $\mathrm{Gpo}(a,b,c)$. The first two moments are

$$\mathrm{E}(X \mid \theta) = \frac{ca}{b} \quad \text{and} \quad \mathrm{Var}(X \mid \theta) = \frac{ca(b+c)}{b^2}.$$

- *Multinomial distribution*: this is a multivariate distribution characterised by the following density:

$$f(\mathbf{x} \mid \boldsymbol{\theta}) = n! \prod_{j=1}^{k} \frac{\pi_j^{x_j}}{x_j!} I\left(\sum_{j=1}^{k} x_j = n\right),$$

with $\mathbf{x} = (x_1,\dots,x_k)$, $x_j \in \mathbb{N}$, $\theta = (n,\boldsymbol{\pi})$ and valid for $\pi_j \in (0,1)$, $\sum_{j=1}^{k} \pi_j = 1$ and $n \in \mathbb{N}$. This is denoted as $\mathrm{Mult}(n,\boldsymbol{\pi})$. The first two moments are

$$\mathrm{E}(X_j \mid \theta) = n\pi_j, \quad \mathrm{Var}(X_j \mid \theta) = n\pi_j(1-\pi_j) \quad \text{and}$$
$$\mathrm{Cov}(X_i, X_j) = -n\pi_i\pi_j$$

for $i \neq j$.

- *Dirichlet-Multinomial distribution*: this is a multivariate distribution characterised by the following density:

$$f(\mathbf{x} \mid \boldsymbol{\theta}) = \frac{\Gamma(n+1)\Gamma(a_0)}{\Gamma(a_0+n)} \prod_{j=1}^{k} \frac{\Gamma(a_j+x_j)}{\Gamma(a_j)\Gamma(x_j)} I\left(\sum_{j=1}^{k} x_j = n\right),$$

with $\mathbf{x} = (x_1,\dots,x_k)$, $x_j \in \mathbb{N}$, $\theta = (n,\mathbf{a})$ where $\mathbf{a} = (a_1,\dots,a_k)$ and valid for $a_j > 0$ and $n \in \mathbb{N}$, with $a_0 = \sum_{j=1}^{k} a_j$. This is denoted as $\mathrm{DMult}(\mathbf{a},n)$. The first two moments are

$$\mathrm{E}(X_j \mid \theta) = n\frac{a_j}{a_0}, \quad \mathrm{Var}(X_j \mid \theta) = \frac{n(n+a_0)a_j(a_0-a_j)}{a_0^2(a_0+1)}$$

$$\text{and} \quad \mathrm{Cov}(X_i, X_j) = -\frac{n(n+a_0)a_i a_j}{a_0^2(a_0+1)}$$

for $i \neq j$.

Continuous Distributions

- *Uniform distribution*: this is characterised by the following density:

$$f(x \mid \theta) = \frac{1}{b-a} I_{(a,b)}(x),$$

with $\theta = (a, b)$ and valid for $a < b \in \mathbb{R}$. We denote it as $\text{Un}(a, b)$. The first two moments are

$$E(X \mid \theta) = \frac{a+b}{2} \quad \text{and} \quad \text{Var}(X \mid \theta) = \frac{(b-a)^2}{12}.$$

- *Beta distribution*: this is characterised by the following density:

$$f(x \mid \theta) = \frac{\Gamma(a+b)}{\Gamma(a)\Gamma(b)} x^{a-1}(1-x)^{b-1} I_{(0,1)}(x),$$

with $\theta = (a, b)$ and valid for $a, b > 0$. We denote it as $\text{Be}(a, b)$. The first two moments are

$$E(X \mid \theta) = \frac{a}{a+b} \quad \text{and} \quad \text{Var}(X \mid \theta) = \frac{ab}{(a+b)^2(a+b+1)}.$$

- *Inverse beta distribution*: this is also known as beta prime or beta of the second kind and is characterised by the following density:

$$f(x \mid \theta) = \frac{\Gamma(a+b)}{\Gamma(a)\Gamma(b)} x^{a-1}(1+x)^{-a-b} I_{(0,\infty)}(x),$$

with $\theta = (a, b)$ and valid for $a, b > 0$. We denote it as $\text{Ibe}(a, b)$. The first two moments are

$$E(X \mid \theta) = \frac{a}{b-1} \quad \text{and} \quad \text{Var}(X \mid \theta) = \frac{a(a+b-1)}{(b-2)(b-1)^2}$$

if $b > 1$ and $b > 2$, respectively.

- *Exponential distribution*: this is characterised by the following density:

$$f(x \mid \theta) = \theta e^{-\theta x} I_{(0,\infty)}(x),$$

valid for $\theta > 0$. We denote it as $\text{Exp}(\theta)$. The first two moments are

$$E(X \mid \theta) = \frac{1}{\theta} \quad \text{and} \quad \text{Var}(X \mid \theta) = \frac{1}{\theta^2}.$$

- *Gamma distribution*: this is characterised by the following density:

$$f(x \mid \theta) = \frac{b^a}{\Gamma(a)} x^{a-1} e^{-bx} I_{(0,\infty)}(x),$$

1.2 Common Probability Distributions

with $\theta = (a,b)$ and valid for $a,b > 0$. We denote it as $\text{Ga}(a,b)$. The first two moments are

$$E(X \mid \theta) = \frac{a}{b} \quad \text{and} \quad \text{Var}(X \mid \theta) = \frac{a}{b^2}.$$

- *Inverse gamma distribution*: this is characterised by the following density:

$$f(x \mid \theta) = \frac{b^a}{\Gamma(a)} x^{-a-1} e^{-b/x} I_{(0,\infty)}(x),$$

with $\theta = (a,b)$ and valid for $a,b > 0$. We denote it as $\text{Iga}(a,b)$. The first two moments are

$$E(X \mid \theta) = \frac{b}{a-1} \quad \text{and} \quad \text{Var}(X \mid \theta) = \frac{b^2}{(a-1)^2(a-2)}$$

if $a > 1$ and $a > 2$, respectively.

- *Gamma-gamma distribution*: this is characterised by the following density:

$$f(x \mid \theta) = \frac{b^a \Gamma(a+c) x^{c-1}}{\Gamma(a)\Gamma(c)(b+x)^{a+c}} I_{(0,\infty)}(x),$$

with $\theta = (a,b,c)$ and valid for $a,b,c > 0$. We denote it as $\text{Gga}(a,b,c)$. The first two moments are

$$E(X \mid \theta) = \frac{cb}{a-1} \quad \text{and} \quad \text{Var}(X \mid \theta) = \frac{b^2 c(a+c-1)}{(a-1)^2(a-2)}$$

if $a > 1$ and $a > 2$, respectively.

- *Normal distribution*: this is characterised by the following density:

$$f(x \mid \theta) = \left(\frac{2\pi}{\tau}\right)^{-1/2} \exp\left\{-\frac{\tau}{2}(x-\mu)^2\right\} I_{\mathbb{R}}(x),$$

with $\theta = (\mu, \tau)$ and valid for $\mu \in \mathbb{R}$ and $\tau > 0$. We denote it as $\text{N}(\mu, \tau)$. The first two moments are

$$E(X \mid \theta) = \mu \quad \text{and} \quad \text{Var}(X \mid \theta) = \frac{1}{\tau}.$$

- *Pareto distribution*: this is characterised by the following density:

$$f(x \mid \theta) = \frac{ab^a}{x^{a+1}} I_{[b,\infty)}(x),$$

with $\theta = (a,b)$ and valid for $a,b > 0$. We denote it as $\text{Pa}(a,b)$. The first two moments are

$$E(X \mid \theta) = \frac{ba}{a-1} \quad \text{and} \quad \text{Var}(X \mid \theta) = \frac{b^2 a}{(a-1)^2(a-2)}$$

if $a > 1$ and $a > 2$, respectively.

- *Inverse Pareto distribution*: this is characterised by the following density:

$$f(x \mid \theta) = ab^a x^{a-1} I_{(0,1/b)}(x),$$

with $\theta = (a,b)$ and valid for $a, b > 0$. We denote it as Ipa(a,b). The first two moments are

$$E(X \mid \theta) = \frac{a}{b(a+1)} \quad \text{and} \quad \text{Var}(X \mid \theta) = \frac{a}{b^2(a+1)^2(a+2)}.$$

- *Student-t*: this is characterised by the following density:

$$f(x \mid \theta) = \frac{\Gamma((\nu+1)/2)}{\Gamma(\nu/2)} \left(\frac{\tau}{\nu\pi}\right)^{1/2} \left(1 + \frac{\tau(x-\mu)^2}{\nu}\right)^{-(\nu+1)/2} I_{\mathbb{R}}(x),$$

with $\theta = (\mu, \tau, \nu)$ and valid for $\mu \in \mathbb{R}$ and $\tau, \nu > 0$. We denote it as St(μ, τ, ν). The first two moments are

$$E(X \mid \theta) = \mu \quad \text{and} \quad \text{Var}(X \mid \theta) = \frac{\nu}{\tau(\nu-2)}$$

if $\nu > 1$ and $\nu > 2$, respectively.

- *Generalised Scaled Student-t*: this is characterised by the following density:

$$f(x \mid \theta) = k(\mu, \tau) \frac{\exp\{\tan^{-1}(x)\tau\mu\}}{(1+x^2)^{1+\tau/2}} I_{\mathbb{R}}(x),$$

with $k(\mu, \tau)$ a normalising constant, $\theta = (\mu, \tau)$ and valid for $\mu \in \mathbb{R}$ and $\tau > 0$. We denote it as GSSt(μ, τ). The first two moments are

$$E(X \mid \theta) = \mu \quad \text{and} \quad \text{Var}(X \mid \theta) = \frac{1+\mu^2}{\tau-1}$$

if $\tau > 1$.

- *Generalised Hyperbolic Secant distribution*: this is characterised by the following density:

$$f(x \mid \theta) = \frac{2^{\tau-2}}{\Gamma(\tau)} \prod_{k=0}^{\infty} \left\{1 + \frac{x^2}{(\tau+2k)^2}\right\}^{-1} \frac{\exp\{\tan^{-1}(\mu)x\}}{(1+\mu^2)^{\tau/2}} I_{\mathbb{R}}(x),$$

with $\theta = (\mu, \tau)$ and valid for $\mu \in \mathbb{R}$ and $\tau > 0$. We denote it as GHS(μ, τ). The first two moments are

$$E(X \mid \theta) = \mu \quad \text{and} \quad \text{Var}(X \mid \theta) = \tau + \mu^2/\tau.$$

- *Dirichlet distribution*: This is a multivariate distribution characterised by the following density:

$$f(\mathbf{x} \mid \boldsymbol{\theta}) = \frac{\Gamma(\sum_{j=1}^{k} \theta_j)}{\prod_{j=1}^{k} \Gamma(\theta_j)} \prod_{j=1}^{k} x_j^{\theta_j - 1} I\left(\sum_{j=1}^{k} x_j = 1\right),$$

where $\mathbf{x} = (x_1, \ldots, x_k)$, $x_j \in (0, 1)$ for $j = 1, \ldots, k$, with $\boldsymbol{\theta} = (\theta_1, \ldots, \theta_k)$ and valid for $\theta_j > 0$ for $j = 1, \ldots, k$. We denote it as $\text{Dir}(\boldsymbol{\theta})$. The first two moments are

$$E(X_j \mid \boldsymbol{\theta}) = \frac{\theta_j}{\sum_{j=1}^{k} \theta_j}, \quad \text{Var}(X_j \mid \boldsymbol{\theta}) = \frac{\theta_j(\sum_{i \neq j} \theta_i)}{(\sum_{i=1}^{k} \theta_i)^2 (\sum_{i=1}^{k} \theta_i + 1)} \quad \text{and}$$

$$\text{Cov}(X_i, X_j) = \frac{-\theta_i \theta_j}{(\sum_{l=1}^{k} \theta_l)^2 (\sum_{l=1}^{k} \theta_l + 1)}$$

for $i \neq j$.
- *Multivariate normal distribution*: This is a multivariate distribution characterised by the following density:

$$f(\mathbf{x} \mid \boldsymbol{\theta}) = (2\pi)^{-p/2} |\mathbf{C}|^{1/2} \exp\left\{-\frac{1}{2}(\mathbf{x} - \boldsymbol{\mu})' \mathbf{C}(\mathbf{x} - \boldsymbol{\mu})\right\} I_{\mathbb{R}^p}(\mathbf{x}),$$

with $\mathbf{x} = (x_1 \ldots, x_p)$, $\boldsymbol{\theta} = (\boldsymbol{\mu}, \mathbf{C})$ and valid for $\boldsymbol{\mu} \in \mathbb{R}^p$ and $\mathbf{C}$ a precision matrix of dimension $p \times p$. We denote it as $N_p(\boldsymbol{\mu}, \mathbf{C})$. The first two moments are

$$E(\mathbf{X} \mid \boldsymbol{\theta}) = \boldsymbol{\mu} \quad \text{and} \quad \text{Var}(\mathbf{X} \mid \boldsymbol{\theta}) = \mathbf{C}^{-1}.$$

In general, we will use a tilde '~' to denote 'distributed as', for example $X \sim \text{Ber}(\theta)$ means that the random variable X has a Bernoulli distribution with parameter θ. We will put an argument in front to denote density, for example $\text{Ber}(x \mid \theta)$ denotes the Bernoulli density. In some cases, to make our statements clear, we will explicitly denote the random variables involved as well as the arguments for densities, for example $f_X(x)$ denotes the density for random variable X evaluated at value x and $f_{X|Y}(x \mid y)$ denotes the conditional density of random variable X given random variable $Y = y$ evaluated at value x. Whenever we can, we will avoid the sub-indexes.

1.3 Moments

In Section 1.1 we used the notation $f(x \mid \theta)$ to denote a parametric density. In this section we remove the explicit dependence on the parameter θ to avoid burdening the notation and simply denote a density as $f(x)$.

Let us recall the definition of moments, marginal and joint. Let X be a real random variable with probability distribution $f(x)$ and let $g(\cdot)$ be a real function. Then the expectation operator E of function g of X is defined as

$$E\{g(X)\} = \int_{\mathbb{R}} g(x) f(x) \, dx \stackrel{\text{or}}{=} \sum_{x \in \mathbb{R}} g(x) f(x) \qquad (1.2)$$

according to whether X is continuous or discrete.

There are two particular cases for g. These are:

- If $g(x) = x$, then $E\{g(X)\} = E(X) = \mu$ and it is called the *mean*. This is also known as the first non-central moment.
- If $g(x) = (x - \mu)^2$, then $E\{g(X)\} = E\{(X - \mu)^2\} = \sigma^2$ and it is called the *variance*. This is also known as the second central moment.

The mean is a measure of central tendency of the values in a random variable, whereas the variance is a measure of dispersion. It is common to consider the squared root of the variance σ, called standard deviation, to measure the dispersion in the same units as the random variable.

Let (X, Y) be a random vector with joint probability distribution $f(x, y)$ and let g be a real function such that $g: \mathbb{R}^2 \to \mathbb{R}$. Then the expectation of g of (X, Y) is defined as

$$E\{g(X,Y)\} = \int_{\mathbb{R}^2} g(x,y) f(x,y) \, dx \, dy \stackrel{\text{or}}{=} \sum_{(x,y) \in \mathbb{R}^2} g(x,y) f(x,y) \qquad (1.3)$$

according to whether the vector (X, Y) is continuous or discrete. Mixture nature of the random variables is also possible with the appropriate changes to expression (1.3). There is one particular case for g that we are interested in. This is

- If $g(x, y) = (x - \mu_X)(y - \mu_Y)$, with μ_X and μ_Y the mean of X and Y, respectively, then $E\{g(X,Y)\} = E\{(X - \mu_X)(Y - \mu_Y)\} = E(XY) - \mu_X \mu_Y = \text{Cov}(X, Y)$ and is called the *covariance*. This is also known as the second cross central moment.

The covariance is a measure of the linear dependence between the two random variables X and Y and it can take any real value. To better interpret the linear dependence, it is customary to compute the covariance of the standardised variables, which produces the correlation, denoted by ρ, and defined as

$$\rho = \text{Corr}(X, Y) = E\left\{\left(\frac{X - \mu_X}{\sigma_X}\right)\left(\frac{Y - \mu_Y}{\sigma_Y}\right)\right\} = \frac{\text{Cov}(X, Y)}{\sigma_X \sigma_Y}.$$

1.3 Moments

The correlation satisfies $\rho \in [-1, 1]$, which makes it easier to determine when the correlation is strong, for values close to -1 or 1, or weak, for values close to zero.

The conditional distributions of X given Y, denoted as $f(x \mid y)$, and the conditional distribution of Y given X, denoted as $f(y \mid x)$ are defined as

$$f(x \mid y) = \frac{f(x,y)}{f(y)} \quad \text{and} \quad f(y \mid x) = \frac{f(x,y)}{f(x)}$$

if the denominators $f(y)$ and $f(x)$ are positive, respectively.

Then, the theorem of total probability states that

$$f(x) = \int_{\mathbb{R}} f(x \mid y) f(y) \, dy \stackrel{\text{or}}{=} \sum_{y \in \mathbb{R}} f(x \mid y) f(y) = \mathrm{E}_Y\{f(x \mid Y)\}, \quad (1.4)$$

that is, we can recover the marginal distribution of X by taking the expected value with respect to Y of the conditional distribution of X given Y. Similarly, $f(y) = \mathrm{E}_X\{f(y \mid X)\}$. With the conditional distributions we can define conditional moments. Let g be a real function; then the expected value of the function g of x with respect to the conditional distribution of X given Y is given by

$$\mathrm{E}\{g(X) \mid y\} = \int_{\mathbb{R}} g(x) f(x \mid y) \, dx \stackrel{\text{or}}{=} \sum_{x \in \mathbb{R}} g(x) f(x \mid y).$$

We note that this conditional expected value $\mathrm{E}\{g(X) \mid y\}$ is a function of the conditioning variable Y. There are two particular cases of interest:

- If $g(x) = x$, then $\mathrm{E}\{g(X) \mid y\} = \mathrm{E}\{X \mid y\} = \mu_{X|y}$ is the conditional mean of X given Y.
- If $g(x) = (x - \mu_{X|y})^2$, then $\mathrm{E}\{g(X) \mid y\} = \mathrm{E}\{(X - \mu_{X|y})^2 \mid y\} = \mathrm{Var}(X \mid y)$ is the conditional variance of X given Y.

Let (X, Y, Z) be a vector of dimension three with joint distribution function $f(x, y, z)$. The conditional distribution of (X, Y) given Z is defined as

$$f(x, y \mid z) = \frac{f(x, y, z)}{f(z)}.$$

Let $g(x, y) = (x - \mu_{X|z})(y - \mu_{Y|z})$; then the conditional covariance of (X, Y) given Z is defined as

$$\mathrm{Cov}(X,Y \mid z) = \mathrm{E}\{g(X,Y) \mid z\} = \mathrm{E}\{(X - \mu_{X|z})(Y - \mu_{Y|z}) \mid z\}$$

$$= \int_{\mathbb{R}^2} (x - \mu_{X|z})(y - \mu_{Y|z}) f(x,y \mid z)\, dx\, dy$$

$$\stackrel{\mathrm{or}}{=} \sum_{(x,y)\in\mathbb{R}^2} (x - \mu_{X|z})(y - \mu_{Y|z}) f(x,y \mid z)$$

$$= \mathrm{E}\{XY \mid z\} - \mathrm{E}(X \mid z)\mathrm{E}(Y \mid z).$$

From the conditional mean, variance and covariance, we can recover the marginal mean, variance and covariance, via the iterative result, which is given as follows.

Proposition 1.1 Mood et al. (1974) *Let (X, Y, Z) be a random vector of dimension three. If the conditional expectations, variances and covariances exist, then*

i). $E(X) = E_Y\{E(X \mid Y)\}$
ii). $\mathrm{Var}(X) = E_Y\{\mathrm{Var}(X \mid Y)\} + \mathrm{Var}_Y\{E(X \mid Y)\}$
iii). $\mathrm{Cov}(X,Y) = E_Z\{\mathrm{Cov}(X,Y \mid Z)\} + \mathrm{Cov}_Z\{E(X \mid Z), E(Y \mid Z)\}$

Proposition 1.1 is the most important result of this section that we will exploit throughout the remaining chapters of the book. Let us present some examples.

Example 1.2 Let (X, N) be a bivariate random vector, whose probability distribution is given by $X \mid N = n \sim \mathrm{Bin}(n, p)$ and $N \sim \mathrm{Bin}(m, q)$. Explicitly, we have

$$f(x \mid n) = \binom{n}{x} p^x (1-p)^{n-x} I_{\{0,1,\ldots,n\}}(x)$$

and

$$f(n) = \binom{m}{n} q^n (1-q)^{m-n} I_{\{0,1,\ldots,m\}}(n).$$

The objective is to find $E(X)$ and $\mathrm{Var}(X)$ in two ways: (a) obtaining the marginal distribution of X using the theorem of total probability (1.4); and (b) using the iterative mean and variance formulae given in Proposition (1.1). For (a) we use the theorem of total probability:

$$f(x) = \mathrm{E}\{f(x \mid N)\} = \sum_n f(x \mid n) f(n)$$

$$= \sum_{n=x}^{m} \frac{n!}{(n-x)! x!} \frac{m!}{(m-n)! n!} p^x (1-p)^{n-x} (1-q)^m \left(\frac{q}{1-q}\right)^n I_{\{0,1,\ldots,m\}}(x).$$

After cancelling some factorials, doing the change of variable $u = n - x$ and completing the combinations, we get

$$f(x) = (pq)^x(1-q)^{m-x}\binom{m}{x}I_{\{0,1,\ldots,m\}}(x)\sum_{u=0}^{m-x}\binom{m-x}{u}\left(\frac{q(1-p)}{1-q}\right)^u.$$

After computing the last sum with Newton's theorem, we get

$$f(x) = \binom{m}{x}(pq)^x(1-pq)^{m-x}I_{\{0,1,\ldots,m\}}(x).$$

Therefore, the marginal distribution of X is another binomial of the form $X \sim \text{Bin}(m,pq)$. In this case, $\text{E}(X) = mpq$ and $\text{Var}(X) = mpq(1-pq)$. Now, for (b) we use the iterative results and obtain that the mean becomes

$$\text{E}(X) = \text{E}\{\text{E}(X \mid N)\} = \text{E}(Np) = p\text{E}(N) = mpq$$

and the variance is

$$\text{Var}(X) = \text{E}\{\text{Var}(X \mid N)\} + \text{Var}\{\text{E}(X \mid N)\} = \text{E}(Np(1-p)) + \text{Var}(Np)$$
$$= p(1-p)\text{E}(N) + p^2\text{Var}(N) = p(1-p)mq + p^2mq(1-q)$$
$$= mpq - mp^2q + mp^2q - mp^2q^2 = mpq(1-pq),$$

which correspond to the previous computed values. As a further illustration, we can compute the conditional distribution $f(n \mid x)$ by using Bayes's theorem (1.1):

$$f(n \mid x) = \frac{\binom{n}{x}p^x(1-p)^{n-x}I_{\{0,1,\ldots,n\}}(x)\binom{m}{n}q^n(1-q)^{m-n}I_{\{0,1,\ldots,m\}}(n)}{\binom{m}{x}(pq)^x(1-pq)^{m-x}I_{\{0,1,\ldots,m\}}(x)}.$$

After re-writing the product of indicator variables in the numerator as $I_{\{x,x+1,\ldots,m\}}(n)I_{\{0,1,\ldots,m\}}(x)$ and cancelling some common terms, we get

$$f(n \mid x) = \binom{m-x}{m-n}\left\{\frac{(1-p)q}{1-pq}\right\}^{n-x}\left(\frac{1-q}{1-pq}\right)^{m-n}I_{\{x,x+1,\ldots,m\}}(n),$$

which can be identified as a shifted binomial, that is, $N - x \mid X = x \sim \text{Bin}\left(m - x, \frac{(1-p)q}{1-pq}\right)$.

Example 1.3 Let (X, N) be a bivariate random vector, whose probability distribution is given by $X \mid N = n \sim \text{Bin}(n,p)$ and $N \sim \text{Po}(\lambda)$. Explicitly, we have

$$f(x \mid n) = \binom{n}{x}p^x(1-p)^{n-x}I_{\{0,1,\ldots,n\}}(x)$$

and
$$f(n) = e^{-\lambda}\frac{\lambda^n}{n!}I_{\{0,1,\dots\}}(n).$$

As in the previous example, the objective is to find E(X) and Var(X) via (a) the marginal distribution of X using the theorem of total probability and (b) using the iterative mean and variance formulae. For (a) we use the theorem of total probability (1.4):

$$f(x) = E\{f(x\mid N)\} = \sum_n f(x\mid n)f(n)$$

$$= \sum_{n=x}^\infty \frac{n!}{(n-x)!x!}\frac{1}{n!}p^x(1-p)^{n-x}e^{-\lambda}\lambda^n I_{\{0,1,\dots\}}(x).$$

After cancelling some factorials, doing the change of variable $u = n - x$, we get

$$f(x) = \frac{e^{-\lambda}}{x!}\left(\frac{p}{1-p}\right)^x I_{\{0,1,\dots\}}(x)\sum_{u=0}^\infty \frac{1}{u!}(\lambda(1-p))^{u+x}.$$

After computing the last sum with Taylor expansion of the exponential function and cancelling some elements, we get

$$f(x) = e^{-\lambda p}\frac{(\lambda p)^x}{x!}I_{\{0,1,\dots\}}(x).$$

Therefore, the marginal distribution of X is a Poisson of the form $X \sim \text{Po}(\lambda p)$. In this case, $E(X) = \lambda p$ and $Var(X) = \lambda p$. Now, for (b) we use Proposition (1.1) and obtain that the mean becomes

$$E(X) = E\{E(X\mid N)\} = E(Np) = pE(N) = p\lambda$$

and the variance is

$$Var(X) = E\{Var(X\mid N)\} + Var\{E(X\mid N)\} = E(Np(1-p)) + Var(Np)$$
$$= p(1-p)E(N) + p^2 Var(N) = p(1-p)\lambda + p^2\lambda$$
$$= p\lambda(1-p+p) = p\lambda.$$

We can compute the conditional distribution $f(n\mid x)$ by using Bayes's theorem (1.1):

$$f(n\mid x) = \frac{f(x\mid n)f(n)}{f(x)}$$
$$= \frac{\binom{n}{x}p^x(1-p)^{n-x}I_{\{0,1,\dots,n\}}(x)e^{-\lambda}\lambda^n\frac{1}{n!}I_{\{0,1,\dots\}}(n)}{e^{-\lambda p}(\lambda p)^n\frac{1}{n!}I_{\{0,1,\dots\}}(x)}.$$

After re-writing the product of indicator variables in the numerator as $I_{\{x,x+1,\ldots\}}(n)I_{\{0,1,\ldots\}}(x)$ and cancelling some common terms, we get

$$f(n \mid x) = e^{-\lambda(1-p)} \frac{\{\lambda(1-p)\}^{n-x}}{(n-x)!} I_{\{x,x+1,\ldots\}}(n),$$

which can be identified as a shifted Poisson, that is, $N - x \mid X = x \sim \text{Po}(\lambda(1-p))$.

1.4 Stochastic Processes

Definition 1.4 A stochastic process, denoted by $\{X(t): t \in \mathbb{T}\}$, is a family or collection of random variables, where t is a parameter that takes values in $\mathbb{T}$. For each t, $X(t)$ is a random variable.

In general, the most common parameter t that indexes a stochastic process is time. In this case $X(t)$ would be the state of the process at time t. However, t could be space in any dimension, for instance in $\mathbb{R}^2$, if $\mathbf{t} = (t_1, t_2)$ then $X(\mathbf{t})$ would be the state of the process at location (t_1, t_2). Sometimes we interchange $X(t)$ with X_t to simplify the notation, avoiding the use of parentheses.

$\mathbb{T}$ is the index set of the process. If $\mathbb{T}$ is enumerable, then $X(t)$ is a process in discrete time, for example $\{X(t): t \in \mathbb{N}\}$. If $\mathbb{T}$ is a non-enumerable subset of $\mathbb{R}$, then $X(t)$ is a process in continuous time, for example $\{X(t): t > 0\}$.

The state of the process is the set of all possible values $X(t) \in \mathbb{X}$ for all $t \in \mathbb{T}$. The state space $\mathbb{X}$ can be discrete or continuous.

One particular type of process of interest is the Markov process. We define it here.

Definition 1.5 A stochastic process $\{X(t): t \in \mathbb{T}\}$ is a Markov process if it satisfies the Markovian property that states given the present, $X(t)$, the values of the future, $X(s)$ for $s > t$, do not depend on the past, $X(u)$ for $u < t$. In notation,

$$P\{X(s) \in A \mid X(u_0) = x_{u_0}, X(u_1) = x_{u_1}, \ldots, X(u_n) = x_{u_n}, X(t) = x_t\}$$
$$= P\{X(s) \in A \mid X(t) = x_t\}$$

for arbitrary $A \subset \mathbb{X}$ and $u_0 < u_1 < \cdots < u_n \leq t < s$.

Another property of interest of stochastic processes is *stationarity*. This is a condition that can be achieved strictly or weakly.

Definition 1.6 A stochastic process $\{X(t): t \in \mathbb{T}\}$ is strictly stationary if it satisfies that for all $n, s, t_1, \ldots, t_n$, the vectors $(X(t_1), \ldots, X(t_n))$ and $(X(t_1 + s), \ldots, X(t_n + s))$ have the same joint distribution.

In other words, Definition 1.6 says that there must be a kind of invariant distribution if we shift the process a specific amount of time. In particular, it must be satisfied that $X(t)$ and $X(t + s)$ must have the same distribution.

A weaker version of stationarity is defined as follows.

Definition 1.7 A stochastic process $\{X(t): t \in \mathbb{T}\}$ is weakly stationary, or second-order stationary, if for $s, t \in \mathbb{T}$, $X(t)$ satisfies the following two conditions:

i). $E\{X(t)\} = \mu$, and
ii). $Cov\{X(t), X(t + s)\} = \sigma(s)$,

That is, the first two moments do not depend on t.

Second-order stationarity, given in Definition 1.7, only requires that the first two moments of the process, mean and variance, remain constant for all times. And after shifting the process, the covariance does not depend on the specific time t, it only depends on the time difference s.

Let us consider a first example.

Example 1.8 Autoregressive process of order 1, $AR(1)$. Let $Z_1, Z_2, \ldots$ be random variables such that $E(Z_t) = 0$ for all t, $Var(Z_t) = \sigma^2$ for $t = 0$, $Var(Z_t) = (1 - \theta^2)\sigma^2$ for $t \geq 1$, and $Cov(Z_t, Z_s) = 0$ for all $t \neq s$. Let

$$X_0 = Z_0 \quad \text{and} \quad X_t = \theta X_{t-1} + Z_t, \text{ for } t \geq 1.$$

Therefore $\{X_t : t \in \mathbb{N}\}$ is an autoregressive process of order 1. If we iterate, we can re-write the process as

$$X_t = \theta(\theta X_{t-2} + Z_{t-1}) + Z_t$$
$$= \theta^2 X_{t-2} + \theta Z_{t-1} + Z_t$$
$$\vdots$$
$$= \sum_{i=0}^{t} \theta^{t-i} Z_i = \sum_{i=0}^{t} \theta^i Z_{t-i}.$$

With this expression we can easily compute the first two moments of the process X_t and the covariance. The mean is

$$E(X_t) = \sum_{i=0}^{t} \theta^{t-i} E(Z_i) = 0.$$

1.4 Stochastic Processes

The variance is

$$\text{Var}(X_t) = \sum_{i=0}^{t} \theta^{2(t-i)} \text{Var}(Z_i) = \theta^{2t} \left\{ \sigma^2 + \sum_{i=1}^{t} \theta^{-2i}(1-\theta^2)\sigma^2 \right\}$$

$$= \sigma^2 \theta^{2t} \left[1 + (1-\theta^2) \left\{ \frac{1-(\theta^{-2})^{t+1}}{1-\theta^{-2}} - 1 \right\} \right]$$

$$= \sigma^2.$$

The covariance is

$$\text{Cov}(X_t, X_{t+s}) = \text{Cov}\left(\sum_{i=0}^{t} \theta^{n-i} Z_i, \sum_{j=0}^{t+s} \theta^{t+s-j} Z_j \right)$$

$$= \sum_{i=0}^{t} \sum_{j=0}^{t+s} \theta^{2t+s-i-j} \text{Cov}(Z_i, Z_j)$$

$$= \sum_{i=0}^{t} \theta^{2t+s-i-j} \text{Var}(Z_i) = \theta^{2t+s} \left\{ \sigma^2 + \sum_{i=1}^{t} \theta^{-2i}(1-\theta^2)\sigma^2 \right\}$$

$$= \sigma^2 \theta^s$$

for $s \geq 0$. Additionally,

$$\text{Corr}(X_t, X_{t+s}) = \theta^s.$$

Since the mean and variance of X_t are constant and the covariance between (X_t, X_{t+s}) only depends on the shift s, $\{X_t\}$ is a second-order stationary process. A further question would be, is $\{X_t\}$ a Markov process? The answer is yes if we add the independence assumption in the $\{Z_t\}$. In such a case

$$f(x_t \mid x_{t-1}, x_{t-2}, \ldots, x_0) = f(x_t \mid x_{t-1}).$$

Note that the version of the autoregressive process of order one presented in Example 1.8 is a finite version of the process, in the sense that it is defined for $t = 0, 1, 2, \ldots, n$ with n finite or infinite. In such a case, to achieve second-order stationarity in $\{X_t\}$, the innovation terms Z_t have different variance for $t = 0$ than for $t \geq 1$. Common specifications of an $AR(1)$ process, for example Chatfield (2003), define the process for non-bounded times, that is, for $t \in \mathbb{Z}$. In such a case we do not need different variances to achieve second order stationarity.

Let us now consider a second example.

Example 1.9 Moving average process of order q, $MA(q)$. Let $Z_1, Z_2, \ldots$ be random variables such that $E(Z_t) = 0$, $Var(Z_t) = \sigma^2$ and $Cov(Z_t, Z_s) = 0$ for all $t \neq s$. Let

$$X_t = \theta_0 Z_t + \theta_1 Z_{t-1} + \cdots + \theta_q Z_{t-q} \quad \text{for } t \in \mathbb{Z}.$$

Therefore $\{X_t : t \in \mathbb{Z}\}$ is a moving average process of order q. We can re-write the process in two different ways

$$X_t = \sum_{i=0}^{q} \theta_i Z_{t-i} = \sum_{j=t-q}^{t} \theta_{t-j} Z_j.$$

With these expressions we can compute the first two moments of the process X_t as well as the covariance. The mean is

$$E(X_t) = E\left(\sum_{i=0}^{q} \theta_i Z_{t-i}\right) = \sum_{i=0}^{q} \theta_i E(Z_{t-i}) = 0.$$

The variance is

$$Var(X_t) = \sum_{i=0}^{q} \theta_i^2 Var(Z_{t-i}) = \sigma^2 \sum_{i=0}^{q} \theta_i^2.$$

The covariance is, for $s \leq q$,

$$Cov(X_t, X_{t+s}) = Cov\left(\sum_{i=t-q}^{t} \theta_{t-i} Z_i, \sum_{j=t+s-q}^{t+s} \theta_{t+s-j} Z_j\right)$$

$$\sum_{i=t-q}^{t} \sum_{j=t+s-q}^{t+s} \theta_{t-i} \theta_{t+s-j} Cov(Z_i, Z_j)$$

$$= \sum_{i=t+s-q}^{t} \theta_{t-i} \theta_{t+s-i} Var(Z_i)$$

$$= \sigma^2 \sum_{i=t+s-q}^{t} \theta_{t-i} \theta_{t+s-i} \quad \text{for } s \leq q.$$

By doing the change of variable $j = i - t - s + q$ in the previous sum, we have

$$Cov(X_t, X_{t+s}) = \sigma^2 \sum_{j=0}^{q-s} \theta_{q-s-j} \theta_{q-j} \quad \text{for } s \leq q$$

and $Cov(X_t, X_{t+s}) = 0$ if $s > q$. Additionally, the correlation becomes

$$Corr(X_t, X_{t+s}) = \frac{\sum_{j=0}^{q-s} \theta_{q-s-j} \theta_{q-j}}{\sum_{i=0}^{q} \theta_i^2} \quad \text{for } s \leq q$$

1.4 Stochastic Processes

and $\text{Corr}(X_t, X_{t+s}) = 0$ if $s > q$. Since the mean and variance of X_t are constants and the covariance does not depend on t, $\{X_t\}$ is a second-order stationary process. But, is $\{X_t\}$ a Markov process? The answer is no because there is no way of writing X_t in terms of X_{t-1} exclusively.

Let us consider a third example of a stochastic process in space instead of time.

Example 1.10 Conditionally autoregressive (CAR) process (Besag, 1974). Let $\{X_i : i = 1, \ldots, n\}$ be a stochastic process such that each random variable X_i is associated to an area i in a region. Let each X_i be defined conditionally on the other areas $j \neq i$ through a normal distribution of the form

$$X_i \mid X_j = x_j, j \neq i \sim \text{N}\left(\sum_j b_{ij} x_j, \tau_i\right).$$

We know that any joint distribution $f(x_1, \ldots, x_n)$ induces well-defined conditional densities $f(x_i \mid x_j, j \neq i)$; However, the converse is not always possible. Brook's lemma (Brook, 1964) states the conditions for obtaining a joint distribution based on its conditional distributions. In this case, it can be proved that

$$f(\mathbf{x}) \propto \exp\left\{-\frac{1}{2}\mathbf{x}'\mathbf{D}(\mathbf{I} - \mathbf{B})\mathbf{x}\right\},$$

where $\mathbf{B} = (b_{ij})$ and $\mathbf{D} = \text{diag}(\tau_1, \ldots, \tau_n)$. For this to be a well-defined joint density, we need the matrix $\mathbf{D}(\mathbf{I} - \mathbf{B})$ to be symmetric. This is satisfied if $b_{ij}\tau_i = b_{ji}\tau_j$ for all i and j. In particular, if $b_{ij} = w_{ij}/w_{i+}$ and $\tau_i = \tau w_{i+}$, where $w_{ij} = I(i \smile j)$ with "$\smile$" denoting neighbour, and $w_{i+} = \sum_j w_{ij}$ is the number of neighbours of area i. In this case the conditional distributions become

$$f(x_i \mid x_j, j \neq i) = \text{N}\left(\sum_j \frac{w_{ij}}{w_{i+}} x_j, \tau w_{i+}\right) \quad (1.5)$$

and the joint distribution is

$$f(x_1, \ldots, x_n) \propto \exp\left\{-\frac{\tau}{2}\mathbf{x}'(\mathbf{D}_w - \mathbf{W})\mathbf{x}\right\}, \quad (1.6)$$

where $\mathbf{W} = (w_{ij})$ and $\mathbf{D}_w = \text{diag}(w_{1+}, \ldots, w_{n+})$. We note that $(\mathbf{D}_w - \mathbf{W})\mathbf{1} = \mathbf{0}$, that is, the precision matrix $(\mathbf{D}_w - \mathbf{W})$ is singular, so the joint distribution (1.6) is improper. Expressions (1.5) and (1.6) define a stochastic process that is known as an *intrinsic CAR* process.

According to Banerjee et al. (2010), the impropriety condition can be corrected by adding an association parameter ρ such that the precision matrix $\mathbf{C} = \mathbf{D}_w - \rho\mathbf{W}$ becomes non-singular. This is achieved if $\rho \in (1/\lambda_{(1)}, 1/\lambda_{(n)})$, where $\lambda_{(1)}$ and $\lambda_{(n)}$ are the minimum and maximum eigenvalues of $\mathbf{D}_w^{-1/2}\mathbf{W}\mathbf{D}_w^{-1/2}$. In this case the conditional distributions become

$$f(x_i \mid x_j, j \neq i) = \mathrm{N}\left(\rho \sum_j \frac{w_{ij}}{w_{i+}} x_j, \tau\, w_{i+}\right)$$

and the joint distribution is $\mathbf{X} \sim \mathrm{N}(\mathbf{0}, \tau(\mathbf{D}_w - \rho\mathbf{W}))$. Alternatively, Cressie (1993) suggested correcting the impropriety condition by considering a parameter $\alpha \in (1/\lambda_{(1)}, 1/\lambda_{(n)})$, where $\lambda_{(1)}$ and $\lambda_{(n)}$ are the minimum and maximum eigenvalues of the adjacency matrix $\mathbf{W}$ and defining a joint distribution $\mathbf{X} \sim \mathrm{N}(\mathbf{0}, \tau(\mathbf{I} - \alpha\mathbf{W}))$. Either of these two latter processes is called a *proper CAR* process.

On the other hand, none of the first two CAR processes are stationary. The first one because it is improper and the second one because $\mathrm{E}(X_i) = 0$ and $\mathrm{Var}(X_i) = 1/(\tau w_{i+})$ and the marginal distribution is not invariant, that is, $X_i \sim \mathrm{N}(0, \tau w_{i+})$. However, the third specification does define a stationary process with invariant distribution $X_i \sim \mathrm{N}(0, \tau)$. Moreover, the three CAR processes satisfy a Markov property in space, because their law only depends on neighbours of the first kind. Therefore intrinsic and proper CAR models are known as *Markov random fields*.

2

Conjugate Models

2.1 Definitions

As we mentioned in Chapter 1, one of the steps in Bayesian inference is to update our prior knowledge by means of Bayes's theorem. In this chapter we will present a family of prior distributions, which after updating with a particular likelihood, the posterior distribution belongs to the same family. These are called *conjugate families*. See, for example, Bernardo and Smith (2000).

Let us establish our ideas in notation. Let $X_1, \ldots, X_n$ be a random sample, that is, a collection of conditionally independent and identically distributed random variables, $X_i \mid \theta \sim f(x \mid \theta)$, for $\theta \in \Theta \subset \mathbb{R}^k$. Let us assume that we specify our prior knowledge on θ through $f(\theta)$. Then, the posterior is given by (1.1), that is, $f(\theta \mid \mathbf{x}) \propto f(\mathbf{x} \mid \theta) f(\theta)$, where "$\propto$" means proportional.

Definition 2.1 (Raiffa and Schlaifer (1961)) If the posterior $f(\theta \mid \mathbf{x})$ belongs to the same family as the prior $f(\theta)$, it is said that $f(\theta)$ is conjugate with respect to the model $f(x \mid \theta)$.

Let us assume that for data $\mathbf{X} = (X_1, \ldots, X_n)$ coming from model $f(x \mid \theta)$ there exists a sufficient statistic of finite dimension, say $S(\mathbf{X})$. In this case the factorisation theorem for sufficient statistics (Halmos and Savage (1949)) establishes that

$$f(\mathbf{x} \mid \theta) = g(s(\mathbf{x}) \mid \theta) h(\mathbf{x}) \qquad (2.1)$$

for an appropriate function g that depends on the sample only through the statistic $S(\mathbf{X})$ and θ, and h a function that only depends on the data. Then a conjugate family is such that

$$f(\theta \mid s_0, n_0) = \frac{f(\mathbf{x} \mid \theta)}{\int_\Theta f(\mathbf{x} \mid \theta) d\theta} = \frac{g(s(\mathbf{x}) \mid \theta)}{\int_\Theta g(s(\mathbf{x}) \mid \theta) d\theta}.$$

Therefore, the conjugate family $f(\theta \mid s_0, n_0)$ is proportional to the likelihood $f(\mathbf{x} \mid \theta)$ as a function of θ. That is, $f(\theta \mid s_0, n_0) \propto f(\mathbf{x} \mid \theta)$.

2.2 Detailed Examples

In this section we present four detailed examples where we start from the sampling model, obtain the sufficient statistic, construct the conjugate family and prove that the posterior belongs to the same class.

Example 2.2 Let $X_1, \ldots, X_n$ be a random sample from $\text{Ber}(\theta)$, for $\theta \in [0,1]$, that is,

$$f(x \mid \theta) = \theta^x (1-\theta)^{1-x} I_{\{0,1\}}(x).$$

Then the likelihood is

$$f(\mathbf{x} \mid \theta) = \prod_{i=1}^{n} \theta^{x_i}(1-\theta)^{1-x_i} I_{\{0,1\}}(x_i) = \theta^{\sum x_i}(1-\theta)^{n-\sum x_i} I_{\{0,1\}^n}(\mathbf{x}).$$

If we take $S(\mathbf{X}) = \sum_{i=1}^n X_i$ as the sufficient statistic for θ, then the conjugate family is

$$f(\theta \mid s_0, n_0) \propto \theta^{s_0}(1-\theta)^{n_0-s_0} I_{[0,1]}(\theta).$$

We identify this as the kernel of a beta distribution of the form

$$f(\theta) = \frac{\Gamma(a+b)}{\Gamma(a)\Gamma(b)} \theta^{a-1}(1-\theta)^{b-1} I_{[0,1]}(\theta),$$

where $a - 1 = s_0$ and $b - 1 = n_0 - s_0$. From this prior we can obtain the posterior as

$$f(\theta \mid \mathbf{x}) \propto \theta^{\sum x_i}(1-\theta)^{n-\sum x_i} \theta^{a-1}(1-\theta)^{b-1} I_{[0,1]}(\theta)$$
$$\propto \theta^{a+\sum x_i - 1}(1-\theta)^{b+n-\sum x_i - 1} I_{[0,1]}(\theta).$$

We identify this as the kernel of another beta distribution,

$$f(\theta \mid \mathbf{x}) = \frac{\Gamma(a_1+b_1)}{\Gamma(a_1)\Gamma(b_1)} \theta^{a_1-1}(1-\theta)^{b_1-1} I_{[0,1]}(\theta),$$

where $a_1 = a + \sum_{i=1}^n x_i$ and $b_1 = b + n - \sum_{i=1}^n x_i$. Therefore, the beta family is conjugate to the Bernoulli for parameter θ. Moreover, the beta family is conjugate to the binomial for parameter θ, that is, if $X_1, \ldots, X_n$ is a sample of independent random variables such that $X_i \mid \theta \sim \text{Bin}(c_i, \theta)$, with known c_i for $i = 1, \ldots, n$ and with prior distribution $\theta \sim \text{Be}(a, b)$, then the posterior distribution is of the form $\theta \mid \mathbf{x} \sim \text{Be}\left(a + \sum_{i=1}^n x_i, b + \sum_{i=1}^n (c_i - x_i)\right)$.

Example 2.3 Let $X_1, \ldots, X_n$ be a random sample from $\text{Un}(0, \theta)$, for $\theta > 0$, that is,

$$f(x \mid \theta) = \frac{1}{\theta} I_{(0,\theta)}(x).$$

Then the likelihood is

$$f(\mathbf{x} \mid \theta) = \prod_{i=1}^{n} \frac{1}{\theta} I_{(0,\theta)}(x_i) = \frac{1}{\theta^n} I_{(0,\theta)^n}(\mathbf{x}).$$

We realise that the indicator is a function of θ, so to re-write it we consider that $0 < x_i < \theta$ for all $i = 1, \ldots, n$, which in terms of the order statistics is equivalent to $0 < x_{(1)} < x_{(2)} < \cdots < x_{(n)} < \theta$. So the likelihood becomes

$$f(\mathbf{x} \mid \theta) = \frac{1}{\theta^n} I_{(x_{(n)}, \infty)}(\theta).$$

If we take $S(\mathbf{X}) = X_{(n)}$, the maximum, as the sufficient statistic for θ, then the conjugate family is

$$f(\theta \mid s_0, n_0) \propto \frac{1}{\theta^{n_0}} I_{(s_0, \infty)}(\theta).$$

To obtain the normalising constant, we solve the integral

$$\int_{s_0}^{\infty} \frac{1}{\theta^{n_0}} d\theta = \left. \frac{\theta^{-n_0+1}}{1-n_0} \right|_{s_0}^{\infty} = \frac{s_0^{-(n_0-1)}}{n_0 - 1},$$

which converges only if $-n_0 + 1 < 0 \iff n_0 > 1$; therefore,

$$f(\theta \mid n_0, s_0) = \frac{(n_0 - 1)s_0^{n_0-1}}{\theta^{n_0}} I_{(s_0, \infty)}(\theta) = \frac{ab^a}{\theta^{a+1}} I_{(b, \infty)}(\theta),$$

where $a = n_0 - 1$ and $b = s_0$. We identify this as the Pareto family. From this prior we can obtain the posterior as

$$f(\theta \mid \mathbf{x}) \propto \frac{1}{\theta^n} I_{(x_{(n)}, \infty)}(\theta) \frac{1}{\theta^{a+1}} I_{(b, \infty)}(\theta) = \frac{1}{\theta^{a+n+1}} I_{(\max\{x_{(n)}, b\}, \infty)}(\theta).$$

We identify this as the kernel of another Pareto distribution of the form

$$f(\theta \mid \mathbf{x}) = \frac{a_1 b_1^{a_1}}{\theta^{a_1+1}} I_{(b_1, \infty)}(\theta),$$

where $a_1 = a + n$ and $b_1 = \max\{b, x_{(n)}\}$. Therefore the Pareto family is conjugate to the uniform on $(0, \theta)$. Moreover, the Pareto family is conjugate to the inverse Pareto (see Section 1.2); that is, if $X_1, \ldots, X_n$ is a sample of independent random variables such that $X_i \mid \theta \sim \text{Ipa}(c_i, 1/\theta)$, with known c_i for $i = 1, \ldots, n$ and with prior distribution $\theta \sim \text{Pa}(a, b)$, then the posterior distribution is of the form $\theta \mid \mathbf{x} \sim \text{Pa}\left(a + \sum_{i=1}^{n} c_i, \max\{b, x_{(n)}\}\right)$.

Example 2.4 Let $X_1, \ldots, X_n$ be a random sample from $\text{Ga}(1, \theta)$, for $\theta > 0$, that is,

$$f(x \mid \theta) = \theta e^{-\theta x} I_{[0, \infty)}(x).$$

Then the likelihood is
$$f(\mathbf{x} \mid \theta) = \theta^n e^{-\theta \sum x_i} I_{[0,\infty)^n}(\mathbf{x}).$$
If we take $S(\mathbf{X}) = \sum_{i=1}^n X_i$ as the sufficient statistic for θ, then the conjugate family is
$$f(\theta \mid s_0, n_0) \propto \theta^{n_0} e^{-\theta s_0} I_{[0,\infty)}(\theta).$$
We identify this as the kernel of a gamma distribution of the form
$$f(\theta) = \frac{b^a}{\Gamma(a)} x^{a-1} e^{-b\theta} I_{[0,\infty)}(\theta),$$
where $a = n_0 + 1$ and $b = s_0$. From this prior we can obtain the posterior as
$$f(\theta \mid \mathbf{x}) \propto \theta^n e^{-\theta \sum x_i} \theta^{a-1} e^{-b\theta} I_{[0,\infty)}(\theta) = \theta^{a+n-1} e^{-(b+\sum x_i)\theta} I_{[0,\infty)}(\theta).$$
We identify this as the kernel of another gamma distribution of the form
$$f(\theta \mid \mathbf{x}) = \frac{b_1^{a_1}}{\Gamma(a_1)} x^{a_1-1} e^{-b_1 \theta} I_{[0,\infty)}(\theta),$$
where $a_1 = a+n$ and $b_1 = b+\sum x_i$. Therefore the gamma family is conjugate with respect to the scale parameter of a gamma. In summary, if $X_1, \ldots, X_n$ is a sample of independent random variables such that $X_i \mid \theta \sim \text{Ga}(c_i, \theta)$ with known c_i for $i = 1, \ldots, n$ and with prior distribution $\theta \sim \text{Ga}(a,b)$, then the posterior distribution is of the form $\theta \mid \mathbf{x} \sim \text{Ga}\left(a + \sum_{i=1}^n c_i, b + \sum_{i=1}^n x_i\right)$.

Example 2.5 Let $X_1, \ldots, X_n$ be a random sample from $\text{Po}(\theta)$, for $\theta > 0$, that is,
$$f(x \mid \theta) = e^{-\theta} \frac{\theta^x}{x!} I_{\{0,1,\ldots\}}(x).$$
Then the likelihood is
$$f(\mathbf{x} \mid \theta) = e^{-n\theta} \frac{\theta^{\sum x_i}}{\prod x_i!} I_{\mathbb{N}^n}(\mathbf{x}).$$
If we take $S(\mathbf{X}) = \sum_{i=1}^n X_i$ as the sufficient statistic for θ, then the conjugate family is
$$f(\theta \mid s_0, n_0) \propto \theta^{s_0} e^{-n_0 \theta} I_{[0,\infty)}(\theta).$$
We identify this as the kernel of a gamma distribution of the form
$$f(\theta) = \frac{b^a}{\Gamma(a)} x^{a-1} e^{-b\theta} I_{[0,\infty)}(\theta),$$
where $a = s_0 + 1$ and $b = n_0$. From this prior we can obtain the posterior as
$$f(\theta \mid \mathbf{x}) \propto e^{-n\theta} \theta^{\sum x_i} \theta^{a-1} e^{-b\theta} I_{[0,\infty)}(\theta) = \theta^{a+\sum x_i - 1} e^{-(b+n)\theta} I_{[0,\infty)}(\theta).$$

We identify this as the kernel of another gamma distribution of the form

$$f(\theta \mid \mathbf{x}) = \frac{b_1^{a_1}}{\Gamma(a_1)} x^{a_1-1} e^{-b_1 \theta} I_{[0,\infty)}(\theta),$$

where $a_1 = a + \sum x_i$ and $b_1 = b + n$. Therefore the gamma family is conjugate with respect to the parameter θ in a Poisson distribution. In summary, if $X_1, \ldots, X_n$ is a sample of independent random variables such that $X_i \mid \theta \sim \text{Po}(c_i \theta)$ with known c_i for $i = 1, \ldots, n$ and with prior distribution $\theta \sim \text{Ga}(a, b)$, then the posterior distribution is of the form $\theta \mid \mathbf{x} \sim \text{Ga}\left(a + \sum_{i=1}^{n} x_i, b + \sum_{i=1}^{n} c_i\right)$.

2.3 Summary of Conjugate Models

The class of conjugate models is not large and includes mainly members of the exponential family. This is because we require a sufficient statistic of finite dimension for constructing the conjugate family. The exception is the Pareto and inverse Pareto case presented in Example 2.3. In each of the following cases we present the sampling distribution (likelihood), the prior distribution, the prior predictive for a single observation and for the sufficient statistic of the sampling model and finally the posterior distribution.

A summary of the most common conjugate families is the following:

- *Beta & Binomial*: The beta distribution is conjugate with respect to a binomial distribution for the success parameter θ. That is, if

$$X_i \mid \theta \stackrel{\text{ind}}{\sim} \text{Bin}(c_i, \theta), \ \forall i = 1, \ldots, n$$

and the prior distribution for θ is

$$\theta \sim \text{Be}(a, b),$$

then, the marginal (prior predictive) distributions for X_i and $\sum_{i=1}^{n} X_i$ are

$$X_i \sim \text{BBin}(a, b, c_i) \quad \text{and} \quad \sum_{i=1}^{n} X_i \sim \text{BBin}\left(a, b, \sum_{i=1}^{n} c_i\right)$$

and the posterior distribution for θ, given the sample $\mathbf{X}$, is

$$\theta \mid \mathbf{x} \sim \text{Be}\left(a + \sum_{i=1}^{n} x_i, b + \sum_{i=1}^{n} (c_i - x_i)\right).$$

- *Beta & Negative binomial*: The beta distribution is also conjugate with respect to a negative binomial distribution for the success parameter θ. That is, if

$$X_i \mid \theta \overset{\text{ind}}{\sim} \text{NB}(c_i, \theta), \ \forall i = 1, \ldots, n$$

and the prior distribution for θ is

$$\theta \sim \text{Be}(a, b),$$

then the marginal (prior predictive) distributions for X_i and $\sum_{i=1}^n X_i$ are

$$X_i \sim \text{BNB}(a, b, c_i) \quad \text{and} \quad \sum_{i=1}^n X_i \sim \text{BNB}\left(a, b, \sum_{i=1}^n c_i\right)$$

and the posterior distribution for θ, given the sample $\mathbf{X}$, is

$$\theta \mid \mathbf{x} \sim \text{Be}\left(a + \sum_{i=1}^n c_i, b + \sum_{i=1}^n x_i\right).$$

- *Inverse Beta & Negative Binomial*: The inverse beta distribution is also conjugate with respect to a negative binomial distribution for the odds associated with the success parameter p, say $\theta = (1-p)/p$. That is, if

$$X_i \mid \theta \overset{\text{ind}}{\sim} \text{NB}(c_i, 1/(\theta+1)), \ \forall i = 1, \ldots, n$$

and the prior distribution for θ is

$$\theta \sim \text{Ibe}(a, b),$$

then the marginal (prior predictive) distributions for X_i and $\sum_{i=1}^n X_i$ are

$$X_i \sim \text{BNB}(b, a, c_i) \quad \text{and} \quad \sum_{i=1}^n X_i \sim \text{BNB}\left(b, a, \sum_{i=1}^n c_i\right)$$

and the posterior distribution for θ, given the sample $\mathbf{X}$, is

$$\theta \mid \mathbf{x} \sim \text{Ibe}\left(a + \sum_{i=1}^n x_i, b + \sum_{i=1}^n c_i\right).$$

- *Gamma & Poisson*: The gamma distribution is conjugate with respect to a Poisson distribution for the rate parameter θ. That is, if

$$X_i \mid \theta \overset{\text{ind}}{\sim} \text{Po}(c_i \theta), \ \forall i = 1, \ldots, n$$

and the prior distribution for θ is

$$\theta \sim \text{Ga}(a, b),$$

2.3 Summary of Conjugate Models

then the marginal (prior predictive) distributions for X_i and $\sum_{i=1}^{n} X_i$ are

$$X_i \sim \text{Gpo}(a,b,c_i) \quad \text{and} \quad \sum_{i=1}^{n} X_i \sim \text{Gpo}\left(a,b,c_i\right)$$

and the posterior distribution for θ, given the sample $\mathbf{X}$, is

$$\theta \mid \mathbf{x} \sim \text{Ga}\left(a + \sum_{i=1}^{n} x_i, b + \sum_{i=1}^{n} c_i\right).$$

- *Gamma & Gamma*: The gamma distribution is also conjugate with respect to a gamma distribution for the rate parameter θ. That is, if

$$X_i \mid \theta \stackrel{\text{ind}}{\sim} \text{Ga}(c_i, \theta), \; \forall i = 1,\ldots,n$$

and the prior distribution for θ is

$$\theta \sim \text{Ga}(a,b),$$

then the marginal (prior predictive) distributions for X_i and $\sum_{i=1}^{n} X_i$ are

$$X_i \sim \text{Gga}(a,b,c_i) \quad \text{and} \quad \sum_{i=1}^{n} X_i \sim \text{Gga}\left(a,b,\sum_{i=1}^{n} c_i\right)$$

and the posterior distribution for θ, given the sample $\mathbf{X}$, is

$$\theta \mid \mathbf{x} \sim \text{Ga}\left(a + \sum_{i=1}^{n} c_i, b + \sum_{i=1}^{n} x_i\right).$$

- *Inverse Gamma & Gamma*: The inverse gamma distribution is also conjugate with respect to a gamma distribution for the scale parameter $1/\theta$. That is, if

$$X_i \mid \theta \stackrel{\text{ind}}{\sim} \text{Ga}(c_i, 1/\theta), \; \forall i = 1,\ldots,n$$

and the prior distribution for θ is

$$\theta \sim \text{Iga}(a,b),$$

then the marginal (prior predictive) distributions for X_i and $\sum_{i=1}^{n} X_i$ are

$$X_i \sim \text{Gga}(a,b,c_i) \quad \text{and} \quad \sum_{i=1}^{n} X_i \sim \text{Gga}\left(a,b,\sum_{i=1}^{n} c_i\right)$$

and the posterior distribution for θ, given the sample $\mathbf{X}$, is

$$\theta \mid \mathbf{x} \sim \text{Iga}\left(a + \sum_{i=1}^{n} c_i, b + \sum_{i=1}^{n} x_i\right).$$

- *Normal & Normal*: The normal distribution is conjugate with respect to a normal distribution for the location parameter θ. That is, if

$$X_i \mid \theta \stackrel{\text{ind}}{\sim} N(\theta, c_i), \ \forall i = 1, \ldots, n$$

and the prior distribution for θ is

$$\theta \sim N(\mu, a),$$

then the marginal (prior predictive) distributions for X_i and $\sum_{i=1}^{n} c_i X_i$ are

$$X_i \sim N\left(\mu, \frac{a c_i}{a + c_i}\right) \quad \& \quad \sum_{i=1}^{n} c_i X_i \sim N\left(\mu \sum_{i=1}^{n} c_i, \frac{a}{\left(\sum_{i=1}^{n} c_i\right)\left(a + \sum_{i=1}^{n} c_i\right)}\right)$$

and the posterior distribution for θ, given the sample $\mathbf{X}$, is

$$\theta \mid \mathbf{x} \sim N\left(\frac{a\mu + \sum_{i=1}^{n} c_i x_i}{a + \sum_{i=1}^{n} c_i}, a + \sum_{i=1}^{n} c_i\right).$$

- *Gamma & Normal*: The gamma distribution is conjugate with respect to a normal distribution for the precision parameter θ. That is, if

$$X_i \mid \theta \stackrel{\text{ind}}{\sim} N(c_i, \theta), \ \forall i = 1, \ldots, n$$

and the prior distribution for θ is

$$\theta \sim Ga(a, b),$$

then the marginal (prior predictive) distributions for X_i and $\frac{1}{2}\sum_{i=1}^{n}(x_i - c_i)^2$ are

$$X_i \sim St\left(c_i, \frac{a}{b}, 2a\right) \quad \text{and} \quad \frac{1}{2}\sum_{i=1}^{n}(X_i - c_i)^2 \sim Gga\left(a, b, \frac{n}{2}\right)$$

and the posterior distribution for θ, given the sample $\mathbf{X}$, is

$$\theta \mid \mathbf{x} \sim Ga\left(a + \frac{n}{2}, b + \frac{1}{2}\sum_{i=1}^{n}(x_i - c_i)^2\right).$$

- *Pareto & Inverse Pareto*: The Pareto distribution is conjugate with respect to an inverse Pareto distribution for the location parameter θ. That is, if

$$X_i \mid \theta \stackrel{\text{ind}}{\sim} Ipa(c_i, 1/\theta), \ \forall i = 1, \ldots, n$$

and the prior distribution for θ is

$$\theta \sim Pa(a, b),$$

2.3 Summary of Conjugate Models

then the marginal (prior predictive) distributions for X_i and $X_{(n)}$ are mixtures of the form

$$f(x_i) = \frac{a}{a+c_i}\text{Ipa}(x_i \mid c_i, 1/b) + \frac{c_i}{a+c_i}\text{Pa}(x_i \mid a, b)$$

and

$$f_{X_{(n)}}(x) = \frac{a}{a+\sum_{i=1}^n c_i}\text{Ipa}\left(x \;\Bigg|\; \sum_{i=1}^n c_i, \frac{1}{b}\right) + \frac{\sum_{i=1}^n c_i}{a+\sum_{i=1}^n c_i}\text{Pa}(x \mid a, b)$$

and the posterior distribution for θ, given the sample $\mathbf{X}$, is

$$\theta \mid \mathbf{x} \sim \text{Pa}\left(a + \sum_{i=1}^n c_i, \max\{b, x_{(n)}\}\right).$$

- *Dirichlet & Multinomial*: The Dirichlet distribution is conjugate with respect to a multinomial distribution for the cell probabilities θ. That is, if

$$\mathbf{X}_i \mid \theta \stackrel{\text{ind}}{\sim} \text{Mult}(c_i, \theta), \; \forall i = 1, \ldots, n$$

and the prior distribution for θ is

$$\theta \sim \text{Dir}(\mathbf{a}),$$

then the marginal (prior predictive) distributions for $\mathbf{X}_i$ and $\sum_{i=1}^n \mathbf{X}_i$ are

$$\mathbf{X}_i \sim \text{DMult}(\mathbf{a}, c_i) \quad \text{and} \quad \sum_{i=1}^n \mathbf{X}_i \sim \text{DMult}\left(\mathbf{a}, \sum_{i=1}^n c_i\right)$$

and the posterior distribution for θ, given the sample $\mathbf{X}$, is

$$\theta \mid \mathbf{x} \sim \text{Dir}\left(\mathbf{a} + \sum_{i=1}^n \mathbf{x}_i\right).$$

- *Multivariate normal & Multivariate normal*: The multivariate normal distribution is conjugate with respect to another multivariate normal distribution for the location parameter θ. That is, if

$$\mathbf{X}_i \mid \theta \stackrel{\text{ind}}{\sim} \text{N}_p(\theta, C_i), \; \forall i = 1, \ldots, n$$

and the prior distribution for θ is

$$\theta \sim \text{N}_p(\mu, A),$$

then the marginal (prior predictive) distributions for $\mathbf{X}_i$ and $\sum_{i=1}^n C_i \mathbf{X}_i$ are

$$\mathbf{X}_i \sim \text{N}_p\left(\mu, (A^{-1} + C_i^{-1})^{-1}\right)$$

and

$$\sum_{i=1}^{n} C_i \mathbf{X}_i \sim \mathrm{N}_p \left(\left(\sum_{i=1}^{n} C_i \right) \mu, \left(I + A^{-1} \sum_{i=1}^{n} C_i \right)^{-1} \left(\sum_{i=1}^{n} C_i \right)^{-1} \right)$$

and the posterior distribution for θ, given the sample $\mathbf{X}$, is

$$\theta \mid \mathbf{x} \sim \mathrm{N}_p \left(\left(A + \sum_{i=1}^{n} C_i \right)^{-1} \left(A\mu + \sum_{i=1}^{n} C_i \mathbf{x}_i \right), A + \sum_{i=1}^{n} C_i \right).$$

This list is not exhaustive; there are other conjugate families not presented here, like the generalised scaled student which is conjugate to the location parameter of the generalised hyperbolic secant distribution. Other quasi-conjugate families are those presented in Examples 1.2 and 1.3, where analytical expressions exist for all conditional and marginal distributions.

The idea of conjugate models can be extended to stochastic processes, like the Dirichlet process commonly used in Bayesian nonparametric statistics which is conjugate to the measure of a multinomial counting process (see Ferguson (1973)), or the gamma process which is conjugate to the rate parameter of a Poisson process. In Chapters 3 and 8 we will present the use of stochastic processes in the construction of dependent sequences.

3
Exchangeable Sequences

3.1 Definitions

There are formal and informal definitions of exchangeability. What is exchangeability informally? If you have data or a sequence of random variables and you drop them and pick them up, the data provide exactly the same information as originally. Exchangeability has to do with order and the order does not matter. For instance, time series data or spatial data, where ordering matters, are not examples of exchangeable sequences.

To understand this idea, let us define exchangeability formally.

Definition 3.1 (de Finetti (1937)) An infinite sequence $X_1, X_2, \ldots$ of random quantities is exchangeable if, for all $n < \infty$, it is satisfied that

$$f_{X_1, X_2, \ldots, X_n}(x_1, x_2, \ldots, x_n) = f_{X_1, X_2, \ldots, X_n}(x_{\pi(1)}, x_{\pi(2)}, \ldots, x_{\pi(n)})$$

for all index permutations π of the set $\{1, 2, \ldots, n\}$.

Definition 3.1 formalises the fact that the order does not matter in an exchangeable sequence. Exchangeability does not imply independence, but the other way around is true; independence is a particular case of exchangeability. So, exchangeability is more general than independence.

Now the definition is clear, but how do we create an exchangeable sequence? To answer this, we refer to another result given by Bruno de Finetti.

Proposition 3.2 (de Finetti (1937)) *If $X_1, X_2, \ldots$ is an infinite exchangeable sequence with probability density f, then there exists a common random quantity $Y \in \mathbb{Y}$ with density $f(y)$ such that for $n < \infty$ we can write*

$$f(x_1, \ldots, x_n) = \int_{\mathbb{Y}} \prod_{i=1}^{n} f(x_i \mid y) f(y) \, dy.$$

33

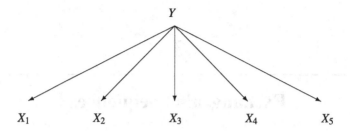

Figure 3.1 Graphical representation of an exchangeable sequence.

Proposition 3.2 tells us that to construct an exchangeable sequence we have to assume that the random quantities of interest, X_i, need to be identically distributed and conditionally independent given a common quantity Y, and the joint law of the X_i is obtained by marginalising with respect to the density of Y, $f(y)$. The result from Proposition 3.2 is represented graphically in Figure 3.1.

Another consequence of Proposition 3.2 is that the marginal distribution for each X_i is the same, so the X_i's have an invariant marginal distribution; however, the X_i are dependent. To see this, let us consider an example.

Example 3.3 Let $X_1, X_2, \ldots$ be a sequence of random variables such that $X_i \mid Y \sim \mathrm{N}(y, \tau)$ are conditionally independent given Y for $i = 1, 2, \ldots$ and $Y \sim \mathrm{N}(\mu_0, \tau_0)$, with known τ. Let us obtain the marginal mean and variance of each X_i and the correlation between any two (X_i, X_j) for $i \neq j = 1, 2, \ldots$. Then, using the iterative expectation formula, we get

$$\mathrm{E}(X_i) = \mathrm{E}\{\mathrm{E}(X_i \mid Y)\} = \mathrm{E}(Y) = \mu_0.$$

Now, using the iterative variance formula, we get

$$\mathrm{Var}(X_i) = \mathrm{E}\{\mathrm{Var}(X_i \mid Y)\} + \mathrm{Var}\{\mathrm{E}(X_i \mid Y)\} = \mathrm{E}\left(\frac{1}{\tau}\right) + \mathrm{Var}(Y) = \frac{1}{\tau} + \frac{1}{\tau_0}.$$

Using properties of normal random variables, the marginal distribution for each X_i is $\mathrm{N}(\mu_0, \tau_0 \tau/(\tau_0 + \tau))$ for all $i = 1, 2, \ldots$. Now, to assess the dependence among variables, we compute the correlation. For the covariance, we also use the iterative formula and the fact that X_i and X_j are conditionally independent given Y; then

$$\mathrm{Cov}(X_i, X_j) = \mathrm{E}\{\mathrm{Cov}(X_i, X_j \mid Y)\} + \mathrm{Cov}\{\mathrm{E}(X_i \mid Y), \mathrm{E}(X_j \mid Y)\}$$
$$= \mathrm{Cov}(Y, Y) = \mathrm{Var}(Y) = \frac{1}{\tau_0}.$$

Finally, the correlation becomes

$$\text{Corr}(X_i, X_j) = \frac{\text{Cov}(X_i, X_j)}{\sqrt{\text{Var}(X_i)\text{Var}(X_j)}} = \frac{1/\tau_0}{(\tau_0 + \tau)/(\tau_0 \tau)} = \frac{\tau}{\tau_0 + \tau}$$

for $i \neq j = 1, 2, \ldots$ Since τ is fixed, let us consider different values of $\tau_0 \in [0, \infty)$,

if $\tau_0 \to 0$ then $\text{Corr}(X_i, X_j) \to 1$, and if $\tau_0 \to \infty$ then $\text{Corr}(X_i, X_j) \to 0$.

Therefore, $\text{Corr}(X_i, X_j) \in [0, 1]$ for $i \neq j = 1, 2, \ldots$.

What we learn from Example 3.3 is that conditionally independent variables, given a common quantity, turn out to be positively correlated, and the strength of the correlation is controlled by the variance (precision) of the common quantity Y. Let us consider another example.

Example 3.4 Let $X_1, X_2, \ldots$ be a sequence of random variables such that $X_i \mid Y \sim \text{Ga}(c, y)$ are conditionally independent given Y for $i = 1, 2, \ldots$ and $Y \sim \text{Ga}(a, b)$, with known c. Let us obtain the marginal mean and variance of each X_i and the correlation between any two (X_i, X_j) for $i = 1, 2, \ldots$. Recall first that $1/Y \sim \text{Iga}(a, b)$, that is, an inverse-gamma distribution (see Chapter 2). Now, using the iterative expectation formula, we get

$$\text{E}(X_i) = \text{E}\{\text{E}(X_i \mid Y)\} = \text{E}\left(\frac{c}{Y}\right) = \frac{cb}{a-1}$$

if $a > 1$. Using the iterative variance formula, we get

$$\text{Var}(X_i) = \text{E}\{\text{Var}(X_i \mid Y)\} + \text{Var}\{\text{E}(X_i \mid Y)\} = \text{E}\left(\frac{c}{Y^2}\right) + \text{Var}\left(\frac{c}{Y}\right)$$

$$= c\left\{\text{Var}\left(\frac{1}{Y}\right) + \text{E}^2\left(\frac{1}{Y}\right)\right\} + c^2 \text{Var}\left(\frac{1}{Y}\right)$$

$$= c\left\{\frac{b^2}{(a-1)(a-2)} + \frac{b^2}{(a-1)^2}\right\} + \frac{c^2 b^2}{(a-1)^2(a-2)}$$

$$= \frac{c b^2(a + c - 1)}{(a-1)^2(a-2)}$$

for $a > 2$. For the covariance, we also use the iterative formula and get

$$\text{Cov}(X_i, X_j) = \text{E}\{\text{Cov}(X_i, X_j \mid Y)\} + \text{Cov}\{\text{E}(X_i \mid Y), \text{E}(X_j \mid Y)\}$$

$$= \text{Cov}\left(\frac{c}{Y}, \frac{c}{Y}\right) = c^2 \text{Var}\left(\frac{1}{Y}\right) = \frac{c^2 b^2}{(a-1)^2(a-2)}.$$

Finally, the correlation becomes

$$\text{Corr}(X_i, X_j) = \frac{\text{Cov}(X_i, X_j)}{\sqrt{\text{Var}(X_i)\text{Var}(X_j)}} = \frac{c}{a+c-1}$$

for $i \neq j = 1, 2, \ldots$. But what is the marginal distribution of each X_i for $i = 1, 2, \ldots$? We would have to solve the integral

$$f(x_i) = \int_{\mathcal{Y}} f(x_i \mid y) f(y) \, dy = \int_0^\infty \frac{y^c}{\Gamma(c)} x_i^{c-1} e^{-yx_i} I_{(0,\infty)}(x_i) \frac{b^a}{\Gamma(a)} y^{a-1} e^{-by} dy$$

$$= \frac{x_i^{c-1} b^a}{\Gamma(c)\Gamma(a)} I_{0,\infty}(x_i) \int_0^\infty y^{a+c-1} e^{(b+x_i)y} dy$$

$$= \frac{b^a \Gamma(a+c)}{\Gamma(c)\Gamma(a)} \frac{x_i^{c-1}}{(b+x_i)^{a+c}} I_{0,\infty}(x_i).$$

This density is proper and is called gamma-gamma and is denoted as Gga(a, b, c).

3.2 Pre-specified Invariant Distributions

In Example 3.4 we started with a conditional gamma for the $X_i \mid Y$ and a marginal gamma for Y and the induced marginal distribution for the X_i turned out to be a gamma-gamma. However, we would like to construct an exchangeable sequence with a desired pre-specified invariant marginal distribution, say $f(x)$. Conditions to achieve this goal are given in the following result.

Proposition 3.5 (Mena and Nieto-Barajas (2010)) *Let X be a random variable with known distribution $f_X(x)$. Let Y be a latent variable with arbitrary conditional distribution $f_{Y\mid X}(y \mid x)$. Then the sequence $X_1, X_2, \ldots$ such that $X_i \mid Y \sim f_{X\mid Y}(x_i \mid y)$ are conditionally independent given Y, imply that marginally $X_i \sim f_X(x_i)$ if and only if $f_{X\mid Y}(x \mid y)$ is the corresponding conditional distribution obtained via Bayes's theorem, that is,*

$$f_{X\mid Y}(x \mid y) = \frac{f_{Y\mid X}(y \mid x) f_X(x)}{f_Y(y)}$$

with $f_Y(y) = \int f_{Y\mid X}(y \mid x) f_X(x) \, dx$ or $f_Y(y) = \sum f_{Y\mid X}(y \mid x) f_X(x)$ according to whether X is continuous or discrete, respectively.

Proof Both densities $f_X(x)$ and $f_{Y\mid X}(y \mid x)$ are given, and assume that Y is continuous, so the marginal distribution of X_i is obtained as the expected value of the conditional distribution with respect to Y, that is,

3.2 Pre-specified Invariant Distributions

$$f_{X_i}(x_i) = \int f_{X_i|Y}(x_i \mid y) f_Y(y) \, dy = \int f_{X|Y}(x_i \mid y) f_Y(y) \, dy$$
$$= \int \frac{f_{Y|X}(y \mid x_i) f_X(x_i)}{f_Y(y)} f_Y(y) \, dy = f_X(x_i) \int f_{Y|X}(y \mid x) \, dy$$
$$= f_X(x_i). \qquad \square$$

Therefore, conditions to achieve a desired invariant distribution in an exchangeable sequence are given by appropriate use of Bayes's theorem. The desired marginal distribution is $f_X(x)$; then with an arbitrary conditional distribution $f_{Y|X}(y \mid x)$, for the latent variable Y we first need to find its marginal distribution via the law of total probability and find the reverse conditional distribution for $X \mid Y$ via Bayes's theorem. We can see this process as having another latent variable Z whose marginal distribution is the desired invariant distribution, that is, $Z \sim f_X(z)$; then conditionally on Z, we define $Y \mid Z \sim f_{Y|X}(y \mid z)$ and finally conditionally on Y, we define the sequence $X_i \mid Y \sim f_{X|Y}(x_i \mid y)$. We can associate the corresponding distributions with those required in a Bayesian updating. That is, the latent variable Z has a distribution that corresponds to the prior, the latent Y has a conditional distribution associated to the likelihood and finally the variables of interest X_i have conditional distributions associated with the posterior. Graphically, the hierarchical representation would be as in Figure 3.2. Let us consider some examples.

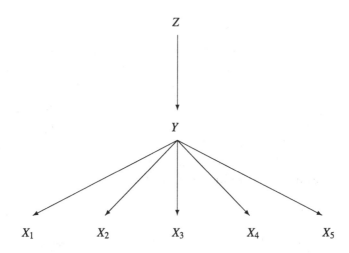

Figure 3.2 Alternative graphical representation of an exchangeable sequence.

Example 3.6 We want to construct a sequence of exchangeable random variables with invariant Ga(a,b) distribution. Then $f(x)$ = Ga($x \mid a,b$). Assume that we choose a latent variable Y with a gamma conditional distribution of the form $f(y \mid x)$ = Ga($y \mid c,x$). Using Bayes's theorem, the conditional distribution of X given Y is another gamma, based on conjugate results from Chapter 2, of the form $f(x \mid y)$ = Ga($x \mid a+c, b+y$), where the marginal distribution of Y is a gamma-gamma distribution $f(y)$ = Gga($y \mid a,b,c$). Therefore, according to Proposition 3.5, $X_1, X_2, \ldots$ are an exchangeable sequence with marginal distribution Ga(a,b) if $X_i \mid Y \sim$ Ga($a+c, b+y$) are conditionally independent given Y for $i = 1, 2, \ldots$ and $Y \sim$ Gga(a,b,c). Alternatively, the gamma-gamma marginal distribution for Y can be written by introducing a new latent variable $Z \sim$ Ga(a,b) and defining $Y \mid Z \sim$ Ga(c,z). Now, what is the dependence induced between any pair (X_i, X_j), $i \neq j$? Using iterative formulae, we get

$$\text{Cov}(X_i, X_j) = E\{\text{Cov}(X_i, X_j \mid Y)\} + \text{Cov}\{E(X_i \mid Y), E(X_j \mid Y)\}$$

$$= \text{Cov}\left(\frac{a+c}{b+Y}, \frac{a+c}{b+Y}\right) = (a+c)^2 \text{Var}\left(\frac{1}{b+Y}\right)$$

$$= \frac{ac}{b^2(a+c+1)}.$$

Since the variance of Ga(a,b) random variables is a/b^2, the correlation becomes

$$\text{Corr}(X_i, X_j) = \frac{c}{a+c+1}.$$

Note that this correlation is different from that of Example 3.4.

In Example 3.6, all computations were possible, due to an appropriate choice of conditional distribution $f(y \mid x)$ which is conjugate to $f(x)$. Let us now consider another example where there is no conjugacy.

Example 3.7 Again, we want to construct an exchangeable sequence with invariant Ga(a,b) distribution. Let $f(x)$ = Ga($x \mid a,b$) and take a conditional distribution for the latent variable Y of the form $f(y \mid x)$ = Un($y \mid 0, x$), which is not conjugate to the gamma. Then, using Bayes's theorem, we obtain the conditional distribution of X, given Y as

$$f(x \mid y) \propto f(y \mid x) f(x) = \frac{1}{x} I_{(0,x)}(y) x^{a-1} e^{-bx} I_{(0,\infty)}(x) = x^{a-2} e^{-bx} I_{(y,\infty)}(x).$$

To obtain the marginal distribution of Y, we do the following:

$$f(y) = \int_y^\infty x^{a-2} e^{-bx} dx = \frac{1}{b^{a-1}} \int_{by}^\infty z^{a-2} e^{-z} dz = \frac{1}{b^{a-1}} \Gamma(a-1, by) I_{(0,\infty)}(y),$$

3.2 Pre-specified Invariant Distributions

where $\Gamma(a, c) = \int_c^\infty x^{a-1} e^{-x} dx$ is the incomplete gamma function. We call this density gamma-uniform and denote it by $\text{Gun}(0, a, b)$. Finally, the conditional distribution of $X \mid Y$ is

$$f(x \mid y) = \frac{b^{a-1}}{\Gamma(a-1, by)} x^{a-2} e^{-bx} I_{(y,\infty)}(x).$$

This distribution has no standard form and can be called truncated gamma and denoted by $\text{Tga}(x \mid a-1, b, by)$. Therefore, according to Proposition 3.5, $X_1, X_2, \ldots$ are an exchangeable sequence with a marginal distribution $\text{Ga}(a, b)$ if $X_i \mid Y \sim \text{Tga}(a-1, b, by)$ are conditionally independent, given Y for $i = 1, 2, \ldots$ and $Y \sim \text{Gun}(0, a, b)$.

Example 3.7 considers a non-conjugate case of $f(y \mid x)$ with respect to the desired marginal distribution $f(x)$. Although the computations were done analytically, this might not be the case for arbitrary $f(y \mid x)$. We suggest, however, to use conjugate models whenever possible to simplify computations.

Let us consider another example for the gamma case but using another conjugate model.

Example 3.8 We want to construct an exchangeable sequence with invariant $\text{Ga}(a, b)$ distribution. Let $f(x) = \text{Ga}(x \mid a, b)$ and now take $f(y \mid x) = \text{Po}(y \mid cx)$ as the conditional distribution of Y given X. Using Bayes's theorem and noting that the Poisson model is conjugate to the gamma, we obtain that the conditional distribution of X given Y is another gamma of the form $f(x \mid y) = \text{Ga}(x \mid a+y, b+c)$, and the marginal distribution for Y is $f(y) = \text{Gpo}(y \mid a, b, c)$, which is a gamma-Poisson density. Therefore, according to Proposition 3.5, $X_1, X_2, \ldots$ are an exchangeable sequence with a $\text{Ga}(a, b)$ marginal distribution if $X_i \mid Y \sim \text{Ga}(a+y, b+c)$ are conditionally independent, given Y for $i = 1, 2, \ldots$ and $Y \sim \text{Gpo}(a, b, c)$. Alternatively, the gamma-Poisson marginal distribution for Y can be written by introducing a latent variable $Z \sim \text{Ga}(a, b)$ and $Y \mid Z \sim \text{Po}(cz)$. In this construction, the dependence measures are easy to compute. The covariance is

$$\text{Cov}(X_i, X_j) = \frac{\text{Var}(Y)}{(b+c)^2} = \frac{1}{(b+c)^2} \frac{ac(b+c)}{b^2} = \frac{ac}{b^2(b+c)},$$

and the correlation becomes

$$\text{Corr}(X_i, X_j) = \frac{c}{b+c}.$$

We have presented four examples involving the gamma density. In Example 3.4 the marginal distribution for the X_i is not gamma, whereas in the last three it is. Examples 3.6 and 3.8 are based on conjugate distributions $f(y \mid x)$ with respect to the desired $f(x)$ and in both cases the conditional distribution of X_i, given Y, is gamma. However, in Example 3.6 the latent variable Y enters the conditional distribution in the second (rate) parameter of the gamma, whereas in Example 3.8 the latent variable Y enters the conditional distribution in the first (shape) parameter of the distribution. Moreover, in the Example 3.6 the correlation is a function of the shape parameter a (and c), whereas in the latter, it is a function of the rate parameter b (and c).

Before proceeding, we show some simulated exchangeable sequences to visualise the effect of parameter c. In the four examples presented here, larger c implies larger correlation. To illustrate, let us consider a sequence with $Ga(1, 1)$ marginal distribution like the one defined in Example 3.6. Figure 3.3 shows five simulated sequences of length $n = 20$ with varying

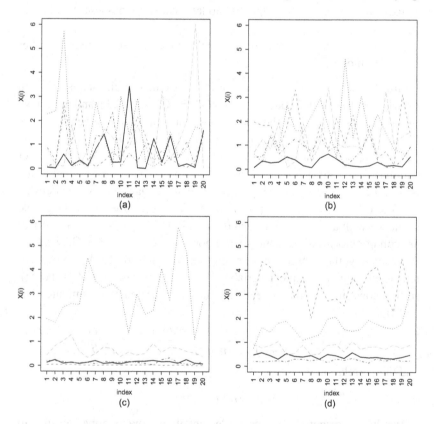

Figure 3.3 Five simulated exchangeable sequences with marginal $Ga(1, 1)$ defined as in Example 3.6. $c = 0$ (**a**), $c = 1$ (**b**), $c = 5$ (**c**) and $c = 20$ (**d**).

values of c. In the top row we took $c = 0$ (panel **(a)**) which implies independent random variables and $c = 1$ (panel **(b)**) which induces a correlation of $1/3 \approx 0.33$. In the bottom row we took $c = 5$ (panel **(c)**) which induces a correlation of $5/7 \approx 0.71$ and $c = 20$ (panel **(d)**) which induces a correlation of $20/22 \approx 0.91$. As we increase the correlation, the sequences start moving less; however, the marginal distribution in all cases is the same.

3.3 Generalisations

One generalisation of the previous ideas is to consider random measures, which are common in the context of Bayesian nonparametrics. Instead of having a latent random variable Y with conditional distribution $f(y \mid x)$, we can have a latent random measure G with conditional distribution $\mathbb{P}(G \mid x)$.

The use of random measures to induce exchangeable sequences with a pre-specified marginal distribution is given in the following proposition.

Proposition 3.9 (Mena and Nieto-Barajas (2010)) *Let X be a random variable with known distribution $f(x)$. Let G be a latent random measure with arbitrary conditional law $\mathbb{P}(G \mid x)$. Then the sequence $X_1, X_2, \ldots$ such that $X_i \mid G \sim G(dx_i)$ are conditionally independent given G and $G \sim \mathbb{P}(G)$ imply that marginally $X_i \sim f(x)$ if and only if*

$$E_G\{G(dx)\} = f(x).$$

Proof We just have to note that the marginal distribution of X_i is given by the expected value of the conditional distribution $G(dx_i)$ with respect to G, that is,

$$E_G\{G(dx)\} = \int G(dx)\mathbb{P}(dG) = F(dx),$$

where $F(\cdot)$ is the cumulative distribution function of density function $f(\cdot)$.
□

The condition given in Proposition 3.9 to achieve a desired marginal distribution via random measures is fortunately satisfied by many random measures proposed in the literature of Bayesian nonparametrics. In order to keep things simple, we will only consider the Dirichlet process introduced by Ferguson (1973) to illustrate.

A Dirichlet process is a random measure G characterised by a scale parameter $c > 0$ and a centring probability measure F. In notation we say $G \sim \mathcal{DP}(c, F)$. For any set $B \subset \mathbb{R}$, the measure assigned to the set B, denoted as $G(B)$, is a beta random variable with parameters $(cF(B), cF(B^c))$; therefore $E\{G(B)\} = F(B)$ and $\text{Var}\{G(B)\} = F(B)F(B^c)/(c+1)$, where $F(B^c) = 1 - F(B)$. If we have data coming from the random measure G, that is, $X_1, \ldots, X_n$ are such that $X_i \mid G \sim G$ and $G \sim \mathcal{DP}(c, F)$, then the joint conditional distribution of the data is $g(x_1, \ldots, x_n)$, where g is the corresponding density associated to G. The expected value of this conditional density was studied by Blackwell and MacQueen (1973) and is usually known as the Pólya urn representation given by

$$X_1 \sim F, \quad X_2 \mid X_1 \sim \frac{c}{c+1}F + \frac{1}{c+1}\widehat{F}_1$$

and in general,

$$X_{i+1} \mid X_1, \ldots, X_i \sim \frac{c}{c+i}F + \frac{i}{c+i}\widehat{F}_i, \qquad (3.1)$$

where $\widehat{F}_i$ is the empirical distribution function of the first i observations.

Now we are in a position to present the example.

Example 3.10 We want to construct an exchangeable sequence with $\text{Ga}(a, b)$ marginal distribution, that is, $f(x) = \text{Ga}(x \mid a, b)$. Let us make G a latent random measure with distribution given by a Dirichlet process with precision parameter $1/c$ and centring measure F, in notation $G \sim \mathcal{DP}(1/c, F)$. Thus, according to Proposition 3.9, $X_1, X_2, \ldots$ are an exchangeable sequence with $\text{Ga}(a, b)$ marginal distribution, if $X_i \mid G \sim G$ are conditionally independent given G for $i = 1, 2, \ldots$ and $G \sim \mathcal{DP}(1/c, F)$ with $F(dx) = \text{Ga}(x \mid a, b) dx$. In this case the dependence measures can be computed but require some extra definitions. The covariance is $\text{Cov}(X_i, X_j) = \text{Var}_G\{E(X \mid G)\}$, where

$$E(X \mid G) = \int xG(dx).$$

Since $E(X \mid G)$ is a functional of the Dirichlet process, it is a random variable whose distribution has been studied by Regazzini et al. (2003) and usually does not have a simple expression. An alternative way of computing the covariance is by marginalising the Dirichlet process first and obtaining the joint distribution of the exchangeable sequence. Let us define $\mu = E(X_i) = a/b$ and $\sigma^2 = \text{Var}(X_i) = a/b^2$. Now, using the Pólya urn (3.1), with c replaced by $1/c$, we compute $E(X_1 X_2) = E\{X_1 E(X_2 \mid X_1)\}$, where $E(X_2 \mid X_1) = \mu/(c+1) + cX_1/(c+1)$, which implies that $E(X_1 X_2) = \mu^2 + \sigma^2 c/(c+1)$. Therefore, $\text{Cov}(X_1, X_2) = \sigma^2 c/(c+1)$ and

$$\text{Corr}(X_1, X_2) = \frac{c}{c+1}.$$

3.4 Applications

We note that this correlation is independent of the marginal distribution of the X_i.

Using a random measure, rather than a random variable, seems to be an easier way to define exchangeable sequences with a pre-specified invariant distribution. The correlation obtained in Example 3.10 is a consequence of using a Dirichlet process as a latent random measure and it is the same regardless of the marginal distribution of the exchangeable sequence.

Another generalisation of the previous ideas is to construct dependent sequences with more flexible dependencies between any pair (X_i, X_j) for $i \neq j$, but losing the invariance of the marginal distribution. For that, we keep considering that our random sequence is defined via conditional independence given a common quantity, but instead of assuming that $X_i \mid Y$ are identically distributed, we can let the parameter c vary according to i, that is c_i. In particular, we consider that $X_i \mid Y \sim f(x_i \mid c_i, y)$ are conditionally independent given Y, for $i = 1, 2, \ldots$, and take $Y \sim f(y)$. In this case the sequence $X_1, X_2, \ldots$ will not be an exchangeable sequence, but will be a dependent sequence. Let us illustrate with an example.

Example 3.11 Let $X_1, X_2, \ldots$ be a sequence of random variables such that $X_i \mid Y \sim \text{Bin}(c_i, y)$ are conditionally independent but not identically distributed, given Y for $i = 1, 2, \ldots$ and $Y \sim \text{Be}(a, b)$. Since the binomial is conjugate to the beta model, the marginal distribution of X_i is easily obtained and is a beta-binomial density denoted as $X_i \sim \text{BBin}(a, b, c_i)$ for $i = 1, 2, \ldots$. Since this marginal distribution depends on c_i, potentially different for each i, this is not an invariant distribution, unless all c_i are the same. However, the X_i are dependent. In particular, the covariance has the form

$$\text{Cov}(X_i, X_j) = \text{Cov}(c_i Y, c_j Y) = c_i c_j \text{Var}(Y) = \frac{c_i c_j a b}{(a+b)^2(a+b+1)}.$$

Since the variance of a beta-binomial random variable is $\text{Var}(X_i) = c_i ab(a + b + c_i)/\{(a + b)^2(a + b + 1)\}$, the correlation becomes

$$\text{Corr}(X_i, X_j) = \sqrt{\frac{c_i c_j}{(a+b+c_i)(a+b+c_j)}}.$$

3.4 Applications

Perhaps the most common application of exchangeable sequences is in the construction of hierarchical models. These are typically used for carrying out meta-analyses. For instance, a COVID-19 study in country i could

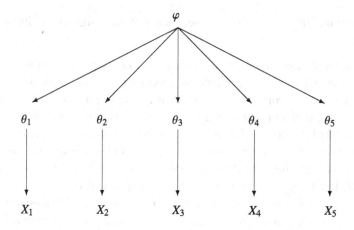

Figure 3.4 Graphical representation of a hierarchical model with $n = 5$.

have a prevalence rate θ_i. Imagine that we want to compare the prevalence rates θ_i for n countries, that is, $i = 1,\ldots,n$. Let X_i be the observable random variable of interest, which in our COVID-19 example could be the number of positive cases in country i. A hierarchical model assumes that the observable random variables X_i are conditionally independent given θ_i for $i = 1,\ldots,n$ and the θ_i since they are all of the same nature, are assumed to be exchangeable. That is, the θ_i are assumed to be conditionally independent, given a common parameter φ, for $i = 1,\ldots,n$ and φ is assigned its own marginal distribution.

In notation, a hierarchical model is defined by a three-level model where in level one, $X_i \mid \theta_i \sim f(x_i \mid \theta_i)$ are conditionally independent, given θ_i for $i = 1,\ldots,n$; in level two, $\theta_i \mid \varphi$ are conditionally independent, given φ for $i = 1,\ldots,n$; and in level three, $\varphi \sim f(\varphi)$. This is depicted in the diagram of Figure 3.4.

Let us consider a specific example.

Example 3.12 Let X_i be the number of COVID-19 cases in country i at a specific period of time. Let P_i be the population size of country i. Let θ_i be the prevalence rate of COVID-19 in country i, for $i = 1,\ldots,n$. To carry out a meta-analysis, we can assume that prevalences are all the same for all countries, that is $\theta_i = \theta$. In this case we can assume a Poisson distribution for the observable random variables of the form $X_i \mid \theta \sim \text{Po}(P_i \theta)$. To carry a Bayesian analysis we need to establish our prior knowledge on the unknown parameters of the model. A typical choice is to consider $\theta \sim \text{Ga}(a,b)$. We call this the *common effects model*.

3.4 Applications 45

Instead of considering that the prevalence rates are all the same, we can assume that they are all different and independent. Model assumptions for this case would be $X_i \mid \theta_i \sim \text{Po}(P_i \theta_i)$ with prior distributions $\theta_i \sim \text{Ga}(a, b)$ independently for $i = 1, \ldots, n$. We call this the *independent effects model*.

An intermediate scenario to carry out a meta-analysis would be to consider a hierarchical model. This would assume that $X_i \mid \theta_i$ are conditionally independent $\text{Po}(P_i \theta_i)$. Now, to define the exchangeable sequence for the θ_i we have two options:

1. The common suggested choice in books for example, Gelman et al. (2013), and also considered in Example 3.4, is to take $\theta_i \mid \varphi \sim \text{Ga}(c, \varphi)$ conditionally independent for $i = 1, \ldots, n$, and $\varphi \sim \text{Ga}(a, b)$. This implies that $\theta_i \sim \text{Gga}(a, b, c)$ marginally for $i = 1, \ldots, n$.
2. The choice suggested in Examples 3.6 and 3.8. Let us consider the former to illustrate. Take $\theta_i \mid \varphi \sim \text{Ga}(a + c, b + \varphi)$ conditionally independent for $i = 1, \ldots, n$, and $\varphi \sim \text{Gga}(a, b, c)$. This implies that $\theta_i \sim \text{Ga}(a, b)$ marginally for $i = 1, \ldots, n$.

Clearly option 2 is more adequate for comparison purposes, with respect to the other two models, than option 1.

We call either of the two options an *exchangeable effects model*.

Since all distributions involved in the previous three models are of standard form, we can run the inference in OpenBUGS (www.mrc-bsu.cam.ac.uk/software/bugs/openbugs/). Bugs code for running these models with the given prior specifications can be found in Tables 3.1, 3.2, and 3.3, respectively.

Let us consider the COVID-19 world data that can be found in Table 1 of the Appendix, updated to 29 November 2023. To illustrate, we only consider

```
model {
#Likelihood
for (i in 1:n) {
x[i] ~ dpois(mu[i])
mu[i] <- pop[i]*theta
}
#Prior
theta ~ dgamma(a,b)
}
```

Table 3.1 *Bugs code for implementing common effects model.*

```
model {
#Likelihood
for (i in 1:n) {
x[i] ~ dpois(mu[i])
mu[i] <- pop[i]*theta[i]
}
#Prior
for (i in 1:n) {
theta[i] ~ dgamma(a,b)
}
}
```

Table 3.2 *Bugs code for implementing independent effects model.*

```
model {
#Likelihood
for (i in 1:n) {
x[i] ~ dpois(mu[i])
mu[i] <- pop[i]*theta[i]
}
#Prior
a1 <- a+c
b1 <- b+phi
for (i in 1:n) {
theta[i] ~ dgamma(a1,b1)
}
phi ~ dgamma(c,d)
d ~ dgamma(a,b)
delta <- a1/b1
}
```

Table 3.3 *Bugs code for implementing exchangeable effects model.*

the top $n = 25$ countries according to their number of cases reported, X_i. Population sizes P_i are also available. Standardised prevalence rates $\widehat{\theta}_i = X_i/P_i$ are shown in panel **(a)** in Figure 3.5. Rates vary from 2.4% in Indonesia to 67.4% in South Korea. This does not necessarily mean that in South Korea almost 70% of the populations have got COVID, since multiple contagions can occur for the same people. The second lowest rate is India with 3.2% and the second highest is Austria with 67.1%.

We implemented our three models to the data, and to avoid numerical problems we scaled the data by dividing by 100,000. Priors were specified

3.4 Applications

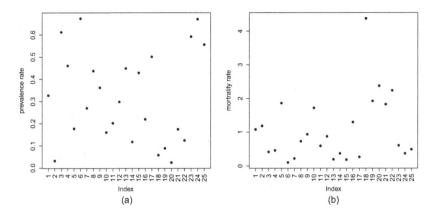

Figure 3.5 COVID-19 data set. Prevalence (a) and mortality (b) rates. The latter are reported in percentages.

with $a = b = 0.01$ and $c = 1$. According to Example 3.4, these values imply a prior correlation on the exchangeable θ_i's of $c/(a + c + 1) = 0.498$. Two chains of the Gibbs samplers were run for 5,000 iterations each with a burn-in of 500 and keeping one of every two iterations. Posterior prevalence point estimates were obtained as the posterior mean (dots) as well as 95% credible intervals obtained with the 2.5% and 97.5% quantiles (lines). These are shown in Figure 3.6.

The two lowest rates correspond to India (ID 2) and Indonesia (ID 20). These two show a very narrow posterior credible interval, which is due to a very large population size. On the other hand, countries with very large posterior credible intervals are Greece (ID 23), Austria (ID 24) and Portugal (ID 25) with the smallest population sizes. For these three countries, interval estimates with the independence model (solid lines) are larger than those obtained with the exchangeable model (dashed lines). Additionally, point rate estimates are smaller for the exchangeable model; this is due to the shrinkage induced by the dependence.

Posterior estimates for the prevalence rate in the common effects model are $E(\theta \mid \text{data}) = 15.6\%$ with a 95% credible interval of $(15.2\%, 16.1\%)$. This interval is very narrow due to the enormous sample size that results from the combination of data from all countries. This is shown in Figure 3.6 after the vertical dotted line. On the other hand, the exchangeable model allows for estimating an overall prevalence rate for all countries as the expected value of the prior conditional distribution for θ_i, that is, $\delta = (a + c)/(b + \varphi)$. Posterior estimates of this quantity are also shown

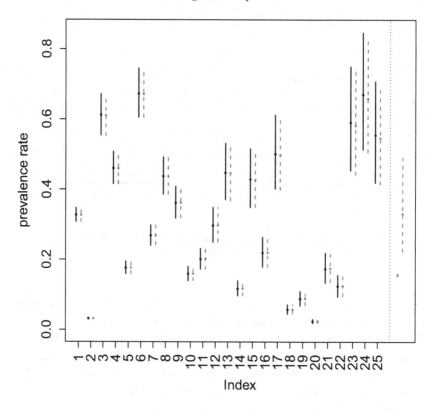

Figure 3.6 COVID-19 data set. Prevalence rates estimates with the three models. Point and 95% credible intervals. Independence model (solid line), hierarchical model (dotted line) and common model (rightmost solid line).

in Figure 3.6 next to the estimates of the common θ. Clearly, this other estimate, of the overall prevalence rate, considers the uncertainty in the different rates θ_i for each country, and is a more realistic representation of the information with a point estimate of 33.13% and 95% credible interval of (22.34%, 49.1%).

Another important aspect of the COVID-19 pandemic is the mortality. Apart from the number of cases, the total number of deaths Y_i is also reported in Table 1 of the Appendix. If we consider the model $Y_i \mid \theta_i, X_i \sim \text{Po}(\theta_i X_i)$, then θ_i becomes the mortality rate. To avoid numerical problems, we scaled the variables, Y_i by 10,000 and X_i by 1,000,000. In panel **(b)** of Figure 3.5, we show the standardised mortality rates $\widehat{\theta_i} = Y_i/X_i$. Values range from 0.10% in South Korea and 0.19% in Taiwan, to 2.38% in Indonesia and 4.37% in Mexico.

3.4 Applications

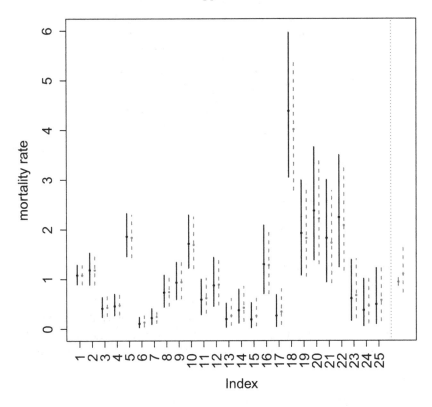

Figure 3.7 COVID-19 data set. Mortality rate estimates with the three models. Point and 95% credible intervals. Independence model (solid line), hierarchical model (dotted line) and common model (rightmost solid line).

We fitted the three previous models assuming common, independent, and exchangeable priors for θ_i. We use the same prior specifications as before with the same setting for the Gibbs samplers. Posterior estimates are shown in Figure 3.7.

Posterior estimate when assuming a common rate is 0.94 (0.86,1.02) deaths for every hundred cases, as compared to an overall rate of 1.1 (0.72,1.65) deaths for every hundred cases estimated with the exchangeable model. These values are shown after the dotted vertical line in Figure 3.7. Since the offsets $X_i/100$ do not take very large values, as compared to the population sizes P_i, the effect of assuming exchangeable priors has more impact on posterior estimates. Values below the overall rate of 1.1 are slightly increased and cases above 1.1 are slightly decreased, with respect to the independence model. This is known as the shrinkage effect. In all cases,

Model	DIC	BIAS	VAR
Prevalence			
Comm	5610	1.69	–
Indep	221.7	1.43×10^{-6}	0.032
Excha	221.5	3.18×10^{-4}	0.031
Mortality			
Comm	374.9	23.38	–
Indep	153.1	0.0001	2.60
Excha	151.1	0.2442	2.30

Table 3.4 *COVID data set. Goodness of fit measures for different models.*

posterior credible intervals with the exchangeable model are narrower than those obtained by the independence model.

Another way of comparing the performance of the models is via goodness of fit measures, like the Deviance Information Criterion (DIC) introduced by Spiegelhalter et al. (2002) or via more predictive measures like the *BIAS* defined as the sum of squared errors or *VAR* defined as the sum of predictive variances. In notation,

$$BIAS = \sum_{i=1}^{n}(X_i - \widehat{X}_i)^2, \quad VAR = \sum_{i=1}^{n} \text{Var}(X_i \mid \mathbf{x}), \quad (3.2)$$

where $\widehat{X}_i = E(X_i \mid \mathbf{x})$ and $\text{Var}(X_i \mid \mathbf{x})$ are the posterior predictive mean and variance, respectively. Of the three indicators the best model is that with the lowest value. To compute the *BIAS*, we use the standardised prevalence (mortality) ratios as the observed data. These indicators are reported in Table 3.4. According to the *DIC*, the worst model is that of common effects model, a high improvement is obtained with the independence effects model and a further small improvement is got from the exchangeable model. In terms of *BIAS*, the best model is the independence effects model because the Bayesian estimates practically coincide with the frequentist standardised ratios. However, in terms of *VAR*, the best model is the exchangeable one reducing the variance in the estimation.

4

Markov Sequences

4.1 Definitions

The aim of this chapter is to show how to construct a discrete-time stochastic process $\{X_i, i \in \mathbb{N}\}$ that satisfies the Markov property. A Markov process was defined in Definition 1.5 in Chapter 1. For a discrete-time process the Markov property states that

$$P(X_i = x_i \mid X_{i-1} = x_{i-1}, \ldots, X_1 = x_1) = P(X_i = x_i \mid X_{i-1} = x_{i-1}).$$

This property can be shown graphically as in Figure 4.1.

An example of a Markov process is the autoregressive process of order one, AR(1), mentioned in Example 1.8. A more general version of it is defined as (e.g. Ross (2009)) $X_1 = \mu + (1/\tau)^{1/2} \xi_1$ and

$$X_{i+1} = \mu(1 - \theta) + \theta X_i + \left(\frac{1 - \theta^2}{\tau}\right)^{1/2} \xi_{i+1} \qquad (4.1)$$

for $i = 1, 2 \ldots$, where $\mu \in \mathbb{R}$, $\tau > 0$, $\theta \in (-1, 1)$ and ξ_i are independent random variables such that $E(\xi_i) = 0$ and $Var(\xi_i) = 1$. However, this process is only partially defined, because only the first two moments of the random variables are available. It is possible to add a normality assumption for each $\xi_i \sim N(0, 1)$, which implies that marginally $X_i \sim N(\mu, \tau)$ such that the process $\{X_i\}$ is strictly stationary with normal invariant distribution and $Corr(X_i, X_{i+s}) = \theta^s$ for $s > 0$.

Instead of explicitly defining the dependence of X_{i+1} in terms of X_i, Nieto-Barajas and Walker (2002) suggested to introduce the dependence via a latent process $\{Y_i, i \in \mathbb{N}\}$, where each Y_i acts as a link between X_i and X_{i+1}. This is described in Figure 4.2, where the arrows mean *conditionally independent*.

Once the latent process $\{Y_i\}$ is marginalised, the process $\{X_i\}$ is a Markov process. Moreover, if for some reason we are interested in the latent process, after marginalising the process $\{X_i\}$, the process $\{Y_i\}$ is also a Markov

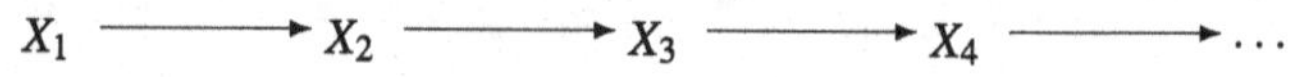

52   *Markov Sequences*

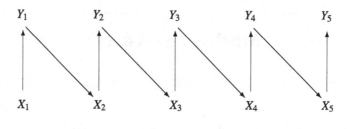

Figure 4.1 Graphical representation of a Markov process.

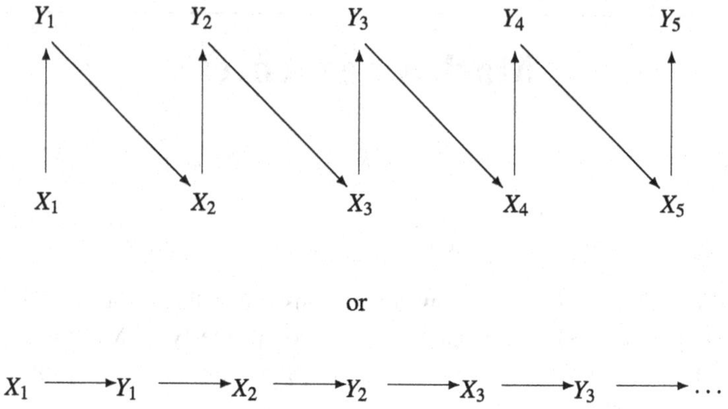

Figure 4.2 Graphical representation of a Markov process via latent variables.

process. In addition to constructing a Markov process our objective is also to define a pre-specified invariant distribution for the X_i, say $f_X(x)$. This is obtained by extending the result given in Proposition 3.5.

Proposition 4.1 *Let X be a random variable with known distribution $f_X(x)$. Let Y be a latent variable with arbitrary conditional distribution $f_{Y|X}(y \mid x)$. Then the sequence $\{X_i\}$ for $i = 1, 2, \ldots$ defined as*

$$X_1 \sim f_X(x_1), \quad Y_i \mid X_i \sim f_{Y|X}(y_i \mid x_i) \quad \text{and} \quad X_{i+1} \mid Y_i \sim f_{X|Y}(x_{i+1} \mid y_i)$$

is a Markov process with invariant distribution $f_X(x)$ if and only if $f_{X|Y}(x \mid y)$ is the corresponding conditional distribution obtained via Bayes's theorem, that is,

$$f_{X|Y}(x \mid y) = \frac{f_{Y|X}(y \mid x) f_X(x)}{f_Y(y)},$$

with $f_Y(y) = \int f_{Y|X}(y \mid x) f_X(x) dx$ or $f_Y(y) = \sum_x f_{Y|X}(y \mid x) f_X(x)$ according to whether X is continuous or discrete, respectively. Moreover, this construction is reversible, that is,

$$Y_i \sim f_Y(y_i), \quad X_i \mid Y_i \sim f_{X|Y}(x_i \mid y_i) \quad Y_{i-1} \mid X_i \sim f_{Y|X}(y_{i-1} \mid x_i),$$

for $i = 1, 2, \ldots$.

4.1 Definitions

Proof Densities $f_X(x)$ and $f_{Y|X}(y \mid x)$ are given. Now, the marginal distribution of Y_1 is the expected value of its conditional distribution, that is,

$$f_{Y_1}(y_1) = \int f_{Y_1|X_1}(y_1 \mid x_1) f_{X_1}(x_1) dx_1$$
$$= \int f_{Y|X}(y_1 \mid x_1) f_X(x_1) dx_1$$
$$= f_Y(y_1),$$

where $f_Y(y)$ is the denominator of Bayes's theorem. To obtain the marginal distribution of X_2, we use again the theorem of total probability, that is,

$$f_{X_2}(x_2) = \int f_{X_2|Y_1}(x_2 \mid y_1) f_{Y_1}(y_1) dy_1$$
$$= \int f_{X|Y}(x_2 \mid y_1) f_Y(y_1) dy_1$$
$$= \int f_{Y|X}(y_1 \mid x_2) f_X(x_2) dy_1$$
$$= f_X(x_2).$$

The proof is concluded by induction. For the reversibility, let us consider the joint distribution of (X_1, Y_1, X_2, Y_2), which is given by

$$f(x_1, y_1, x_2, y_2) = f_{X_1}(x_1) f_{Y_1|X_1}(y_1 \mid x_1) f_{X_2|Y_1}(x_2 \mid y_1) f_{Y_2|X_2}(y_2 \mid x_2)$$
$$= f_X(x_1) f_{Y|X}(y_1 \mid x_1) \frac{f_{Y|X}(y_1 \mid x_2) f_X(x_2)}{f_Y(y_1)} \frac{f_{X|Y}(x_2 \mid y_2) f_Y(y_2)}{f_X(x_2)}.$$

Cancelling the marginal distributions of X_2 and rearranging the terms, we get

$$f(x_1, y_1, x_2, y_2) = f_Y(y_2) f_{X|Y}(x_2 \mid y_2) f_{Y|X}(y_1 \mid x_2) \frac{f_{Y|X}(y_1 \mid x_1) f_X(x_1)}{f_Y(y_1)}$$
$$= f_{Y_2}(y_2) f_{X_2|Y_2}(x_2 \mid y_2) f_{Y_1|X_2}(y_1 \mid x_2) f_{X_1|Y_1}(x_1 \mid y_1),$$

which is the joint distribution of the same vector written in reverse order, namely (Y_2, X_2, Y_1, X_1). This completes the proof. □

Therefore, the way of constructing a Markov process with a pre-specified marginal distribution $f_X(x)$ is to choose arbitrary conditional distribution for the latent variables $f_{Y|X}(y \mid x)$ and define the reverse conditional $f_{X|Y}(x \mid y)$ via Bayes's theorem and use them as building blocks. We can think of these as the prior, likelihood and posterior distributions, respectively. As discussed in Chapter 2, Bayesian learning has closed form expressions for conjugate families, so we will stick to them.

The reversibility of the process stated in Proposition 4.1 will be helpful in the characterisation of the dependence in the sequences $\{X_i\}$ and $\{Y_i\}$, which will be given by the correlation. In particular, we will obtain the covariance between (Y_{i-1}, Y_i) using expressions for $Y_{i-1} \leftarrow X_{i-1} \rightarrow Y_i$ and conditioning on X_{i-1}, and the covariance between (X_i, X_{i+1}) using expressions for $X_i \leftarrow Y_i \rightarrow X_{i+1}$ and conditioning on Y_i.

4.2 Examples

Let us start with an example where we use the beta as conjugate family for the success parameter of the binomial distribution.

Example 4.2 Our building blocks are the beta, binomial and beta distributions. Let

$$X_1 \sim \text{Be}(a, b), \quad Y_i \mid X_i \sim \text{Bin}(c_i, x_i)$$

and

$$X_{i+1} \mid Y_i \sim \text{Be}(a + y_i, b + c_i - y_i)$$

with $a, b > 0$ and $c_i \in \mathbb{N}$ for $i = 1, 2, \ldots$. Since this is a conjugate model, the marginal distributions can be easily computed as

$$Y_i \sim \text{BBin}(a, b, c_i) \quad \text{and} \quad X_i \sim \text{Be}(a, b).$$

Therefore $\{Y_i\}$ is a Markov process with $\text{BBin}(a, b, c_i)$ marginals, which is not an invariant distribution unless $c_i = c$ for all $i = 1, 2, \ldots$, and $\{X_i\}$ is another Markov process with $\text{Be}(a, b)$ marginals, which are invariant. Using the reversibility of the process, we obtain that

$$X_i \mid Y_i \sim \text{Be}(a + y_i, b + c_i - y_i) \quad \text{and} \quad Y_{i-1} \mid X_i \sim \text{Bin}(c_{i-1}, x_i).$$

Therefore, the covariance in the process $\{Y_i\}$ is

$$\text{Cov}(Y_{i-1}, Y_i) = \text{E}\{\text{Cov}(Y_{i-1}, Y_i \mid X_i)\} + \text{Cov}\{\text{E}(Y_{i-1} \mid X_i), \text{E}(Y_i \mid X_i)\},$$

where the first term is zero due to conditional independence, then

$$\text{Cov}(Y_{i-1}, Y_i) = \text{Cov}(c_{i-1} X_i, c_i X_i) = c_{i-1} c_i \text{Var}(X_i) = \frac{c_{i-1} c_i ab}{(a+b)^2(a+b+1)}.$$

Since $\text{Var}(Y_i) = c_i ab(a+b+c_i)/(a+b)^2/(a+b+c_i)$, the correlation becomes

$$\text{Corr}(Y_{i-1}, Y_i) = \sqrt{\frac{c_{i-1} c_i}{(a+b+c_{i-1})(a+b+c_i)}} \stackrel{c_i=c}{=} \frac{c}{a+b+c}.$$

4.2 Examples

Now, the covariance in the process $\{X_i\}$ is

$$\text{Cov}(X_i, X_{i+1}) = \text{Cov}\left(\frac{a+Y_i}{a+b+c_i}, \frac{a+Y_i}{a+b+c_i}\right) = \frac{\text{Var}(Y_i)}{(a+b+c_i)^2}$$

$$= \frac{c_i ab(a+b+c_i)}{(a+b)^2(a+b+1)(a+b+c_i)^2}.$$

Since $\text{Var}(X_i) = ab/\{(a+b)^2(a+b+1)\}$, the correlation becomes

$$\text{Corr}(X_i, X_{i+1}) = \frac{c_i}{a+b+c_i} \stackrel{c_i=c}{=} \frac{c}{a+b+c}.$$

In general, the correlations for the two processes are different, however, if $c_i = c$ for all $i = 1, 2, \ldots$, then we have two Markov processes with $\text{BBin}(a, b, c)$ and $\text{Be}(a, b)$ invariant distributions and with exactly the same correlation structure.

In Example 4.2 we constructed two Markov processes with beta-binomial and beta marginal distributions. But is it possible to construct a Markov process, using these ideas, with binomial marginal distribution? The answer is yes and for that we will use distributions in Example 1.2, where we showed that the binomial distribution is quasi-conjugate to the number of trials in another binomial distribution.

Example 4.3 Our building blocks are the binomial, binomial and shifted binomial distributions. Let

$$X_1 \sim \text{Bin}(n, p), \quad Y_i \mid X_i \sim \text{Bin}(x_i, c_i),$$

with $c_i \in (0, 1)$ and

$$X_{i+1} - y_i \mid Y_i \sim \text{Bin}\left(n - y_i, \frac{p - c_i p}{1 - c_i p}\right),$$

for $i = 1, 2, \ldots$. Although this is not a conjugate model, the corresponding conditional distributions can be obtained analytically. Using results from Example 1.2, the marginal distributions are

$$Y_i \sim \text{Bin}(n, c_i p) \quad \text{and} \quad X_i \sim \text{Bin}(n, p).$$

Therefore $\{Y_i\}$ is a Markov process with $\text{Bin}(n, c_i p)$ marginals, which are not invariant unless $c_i = c$ for all $i = 1, 2, \ldots$, and $\{X_i\}$ is another Markov process with $\text{Bin}(n, p)$ marginals, which are invariant. Using the reversibility of the process, we obtain that

$$X_i - y_i \mid Y_i \sim \text{Bin}\left(n - y_i, \frac{p - c_i p}{1 - c_i p}\right) \quad \text{and} \quad Y_{i-1} \mid X_i \sim \text{Bin}(x_i, c_{i-1}).$$

Therefore, the covariance in the process $\{Y_i\}$ is

$$\text{Cov}(Y_{i-1}, Y_i) = \text{Cov}(c_{i-1}X_i, c_iX_i) = c_{i-1}c_i\text{Var}(X_i) = c_{i-1}c_inp(1-p).$$

Since $\text{Var}(Y_i) = nc_ip(1-c_ip)$, the correlation becomes

$$\text{Corr}(Y_{i-1}, Y_i) = \sqrt{\frac{c_{i-1}c_i}{(1-c_{i-1}p)(1-c_ip)}}(1-p) \stackrel{c_i=c}{=} \frac{c(1-p)}{1-cp}.$$

Now, the covariance of the process $\{X_i\}$, after factorising Y_i on both sides, is

$$\text{Cov}(X_i, X_{i+1}) = \text{Cov}\left\{(n-Y_i)\left(\frac{p-c_ip}{1-c_ip}\right) + Y_i, (n-Y_i)\left(\frac{p-c_ip}{1-c_ip}\right) + Y_i\right\}$$

$$= \left\{1 - \left(\frac{p-c_ip}{1-c_ip}\right)\right\}^2 \text{Var}(Y_i) = \left(\frac{1-p}{1-c_ip}\right)^2 nc_ip(1-c_ip).$$

Since $\text{Var}(X_i) = np(1-p)$, the correlation becomes

$$\text{Corr}(X_i, X_{i+1}) = \frac{c_i(1-p)}{1-c_ip} \stackrel{c_i=c}{=} \frac{c(1-p)}{1-cp}.$$

Let us now show another example, but using the gamma as the conjugate family for the rate parameter in a Poisson likelihood.

Example 4.4 Our building blocks are the gamma, Poisson and gamma distributions. Let

$$X_1 \sim \text{Ga}(a, b), \quad Y_i \mid X_i \sim \text{Po}(c_ix_i)$$

and

$$X_{i+1} \mid Y_i \sim \text{Ga}(a + y_i, b + c_i)$$

with $a, b, c_i > 0$ for $i = 1, 2, \ldots$. Taking advantage of conjugacy, the marginal distributions are

$$Y_i \sim \text{Gpo}(a, b, c_i) \quad \text{and} \quad X_i \sim \text{Ga}(a, b).$$

Therefore, $\{Y_i\}$ is a Markov process with $\text{Gpo}(a, b, c_i)$ marginals, which is not an invariant distribution unless $c_i = c$ for all $i = 1, 2, \ldots$, and $\{X_i\}$ is another Markov process with $\text{Ga}(a, b)$ marginals, which are invariant. Using the reversibility of the process, we obtain that

$$X_i \mid Y_i \sim \text{Ga}(a + y_i, b + c_i) \quad \text{and} \quad Y_{i-1} \mid X_i \sim \text{Po}(c_{i-1}x_i).$$

Therefore, the covariance in the process $\{Y_i\}$ is

$$\text{Cov}(Y_{i-1}, Y_i) = \text{Cov}(c_{i-1}X_i, c_iX_i) = c_{i-1}c_i\text{Var}(X_i) = \frac{c_{i-1}c_ia}{b^2}.$$

4.2 Examples

Since $\text{Var}(Y_i) = c_i(b + c_i)a/b^2$, the correlation becomes

$$\text{Corr}(Y_{i-1}, Y_i) = \sqrt{\frac{c_{i-1}c_i}{(b + c_{i-1})(b + c_i)}} \stackrel{c_i = c}{=} \frac{c}{b + c}.$$

Now, the covariance in the process $\{X_i\}$ is

$$\text{Cov}(X_i, X_{i+1}) = \text{Cov}\left(\frac{a + Y_i}{b + c_i}, \frac{a + Y_i}{b + c_i}\right) = \frac{\text{Var}(Y_i)}{(b + c_i)^2} = \frac{c_i a(b + c_i)}{b^2(b + c_i)^2}.$$

Since $\text{Var}(X_i) = a/b^2$, the correlation becomes

$$\text{Corr}(X_i, X_{i+1}) = \frac{c_i}{b + c_i} \stackrel{c_i = c}{=} \frac{c}{b + c}.$$

Again, the correlations for the two processes are different; however, if $c_i = c$ for all $i = 1, 2, \ldots$ then we have two Markov processes with $\text{Gpo}(a, b, c)$ and $\text{Ga}(a, b)$ invariant distributions, respectively, and with exactly the same correlation structure.

How do we construct a process with Poisson marginal distributions? This is given in the following example. We rely on the fact that the Poisson family is quasi-conjugate with respect to the number-of-successes parameter in a binomial distribution, as shown in Example 1.3.

Example 4.5 Our building blocks are the Poisson, binomial and shifted Poisson distributions. Let

$$X_1 \sim \text{Po}(\mu), \quad Y_i \mid X_i \sim \text{Bin}(x_i, c_i),$$

with $c_i \in (0, 1)$ and

$$X_{i+1} - y_i \mid Y_i \sim \text{Po}(\mu(1 - c_i)),$$

for $i = 1, 2, \ldots$. This is not a conjugate model, but conditional distributions can all be computed explicitly. As shown in Example 1.3, the marginal distributions are

$$Y_i \sim \text{Po}(\mu c_i) \quad \text{and} \quad X_i \sim \text{Po}(\mu).$$

Therefore, $\{Y_i\}$ is a Markov process with $\text{Po}(\mu c_i)$ marginals, which are not invariant unless $c_i = c$ for all $i = 1, 2, \ldots$, and $\{X_i\}$ is another Markov process with $\text{Po}(\mu)$ marginals, which are invariant. Using the reversibility of the process, we obtain that

$$X_i - y_i \mid Y_i \sim \text{Po}(\mu(1 - c_i)) \quad \text{and} \quad Y_{i-1} \mid X_i \sim \text{Bin}(x_i, c_{i-1}).$$

Therefore, the covariance of the process $\{Y_i\}$ is

$$\text{Cov}(Y_{i-1}, Y_i) = \text{Cov}(c_{i-1}X_i, c_iX_i) = c_{i-1}c_i\text{Var}(X_i) = c_{i-1}c_i\mu.$$

Since $\text{Var}(Y_i) = c_i\mu$, the correlation becomes

$$\text{Corr}(Y_{i-1}, Y_i) = \sqrt{c_{i-1}c_i} \stackrel{c_i=c}{=} c.$$

Now, the covariance of the process $\{X_i\}$ is

$$\text{Cov}(X_i, X_{i+1}) = \text{Cov}\,(\mu(1-c_i) + Y_i, \mu(1-c_{i+1}) + Y_i) = \text{Var}(Y_i) = c_i\mu.$$

Since $\text{Var}(X_i) = \mu$, the correlation becomes

$$\text{Corr}(X_i, X_{i+1}) = c_i \stackrel{c_i=c}{=} c.$$

We now show how to construct a Markov process with normal invariant distribution, using the fact that the normal distribution is conjugate with respect to the location parameter in another normal likelihood.

Example 4.6 Our building blocks are normal, normal and normal distributions. Let

$$X_1 \sim \text{N}(\mu, \tau), \quad Y_i \mid X_i \sim \text{N}(x_i, c_i)$$

and

$$X_{i+1} \mid Y_i \sim \text{N}\left(\frac{\tau\mu + c_i y_i}{\tau + c_i}, \tau + c_i\right)$$

with $\mu \in \mathbb{R}$, $\tau > 0$ and $c_i > 0$ for $i = 1, 2, \ldots$. Taking advantage of conjugacy, the marginal distributions are

$$Y_i \sim \text{N}\left(\mu, \frac{\tau c_i}{\tau + c_i}\right) \quad \text{and} \quad X_i \sim \text{N}(\mu, \tau).$$

Therefore, $\{Y_i\}$ is a Markov process with $\text{N}(\mu, \tau c_i/(\tau + c_i))$ marginals, which is not an invariant distribution unless $c_i = c$ for all $i = 1, 2, \ldots$, and $\{X_i\}$ is another Markov process with $\text{N}(\mu, \tau)$ marginals, which are invariant. Using the reversibility of the process, we obtain that

$$X_i \mid Y_i \sim \text{N}\left(\frac{\tau\mu + c_i y_i}{\tau + c_i}, \tau + c_i\right) \quad \text{and} \quad Y_{i-1} \mid X_i \sim \text{N}(x_i, c_{i-1}).$$

Therefore, the covariance in the process $\{Y_i\}$ is

$$\text{Cov}(Y_{i-1}, Y_i) = \text{Cov}(X_i, X_i) = \text{Var}(X_i) = \frac{1}{\tau}.$$

Since $\text{Var}(Y_i) = (\tau + c_i)/(\tau c_i)$, the correlation becomes

$$\text{Corr}(Y_{i-1}, Y_i) = \sqrt{\frac{c_{i-1}c_i}{(\tau + c_{i-1})(\tau + c_i)}} \stackrel{c_i=c}{=} \frac{c}{\tau + c}.$$

4.2 Examples

Now, the covariance in the process $\{X_i\}$ is

$$\text{Cov}(X_i, X_{i+1}) = \text{Cov}\left(\frac{\tau\mu + c_i Y_i}{\tau + c_i}, \frac{\tau\mu + c_i Y_i}{\tau + c_i}\right) = \frac{c_i^2}{(\tau + c_i)^2}\text{Var}(Y_i) = \frac{c_i}{\tau(\tau + c_i)}.$$

Since $\text{Var}(X_i) = 1/\tau$, the correlation becomes

$$\text{Corr}(X_i, X_{i+1}) = \frac{c_i}{\tau + c_i} \stackrel{c_i=c}{=} \frac{c}{\tau + c}.$$

Again, the correlations for the two processes are different; however, if $c_i = c$ for all $i = 1, 2, \ldots$ then we have two Markov processes with $N(\mu, \tau c/(\tau + c))$ and $N(\mu, \tau)$ invariant distributions and with exactly the same correlation structure.

Up to now we have only computed the dependence measure of order (lag) one, but we can also compute dependency measures of other orders, say between X_i and X_{i+2}. For this we would need to re-write the process as $X_i \leftarrow Y_i \leftarrow X_{i+1} \rightarrow Y_{i+1} \rightarrow X_{i+2}$ so that we can doubly condition, first on (Y_i, Y_{i+1}) and finally on X_{i+1}. That is,

$$\begin{aligned}\text{Cov}(X_i, X_{i+2}) &= \text{Cov}\left(\frac{\tau\mu + c_i Y_i}{\tau + c_i}, \frac{\tau\mu + c_{i+1} Y_{i+1}}{\tau + c_{i+1}}\right) \\ &= \frac{c_i c_{i+1}}{(\tau + c_i)(\tau + c_{i+1})}\text{Cov}(Y_i, Y_{i+1}) \\ &= \frac{c_i c_{i+1}}{(\tau + c_i)(\tau + c_{i+1})}\text{Cov}(X_{i+1}, X_{i+1}) \\ &= \frac{c_i c_{i+1}}{(\tau + c_i)(\tau + c_{i+1})\tau}.\end{aligned}$$

Since $V(X_i) = 1/\tau$, the correlation becomes

$$\text{Corr}(X_i, X_{i+2}) = \frac{c_i c_{i+1}}{(\tau + c_i)(\tau + c_{i+1})} \stackrel{c_i=c}{=} \frac{c^2}{(\tau + c)^2}.$$

So, in general, for $s > 0$ we deduce that

$$\text{Corr}(X_i, X_{i+s}) = \prod_{j=0}^{s-1} \frac{c_{i+j}}{(\tau + c_{i+j})} \stackrel{c_i=c}{=} \left(\frac{c}{\tau + c}\right)^s.$$

This last expression, for $c_i = c$, is similar to the correlation induced by the $AR(1)$ process defined in Section 4.1. In order to establish an equivalence we re-write our Markov construction in the following way. Given Y_i,

$$X_{i+1} = \frac{\tau\mu}{\tau + c_i} + \frac{c_i}{\tau + c_i}Y_i + \frac{1}{(\tau + c_i)^{1/2}}\epsilon_{i+1}$$

with $\epsilon_i \sim N(0,1)$. Moreover, given X_i, we also have $Y_i = X_i + c_i^{-1/2}\varepsilon_i$ with $\varepsilon_i \sim N(0,1)$, where ϵ_{i+1} and ε_i are independent. Therefore,

$$X_{i+1} = \frac{\tau\mu}{\tau + c_i} + \frac{c_i}{\tau + c_i}X_i + \left(\frac{c_i}{\tau + c_i}\right)c_i^{-1/2}\varepsilon_i + \frac{1}{(\tau + c_i)^{1/2}}\epsilon_{i+1}$$

$$= \frac{\tau\mu}{\tau + c_i} + \frac{c_i}{\tau + c_i}X_i + \frac{(\tau + 2c_i)^{1/2}}{\tau + c_i}\xi_{i+1}$$

$$= \mu(1 - \theta_i) + \theta_i X_i + \left(\frac{1 - \theta_i^2}{\tau}\right)^{1/2}\xi_{i+1}, \qquad (4.2)$$

where $\theta_i = c_i/(\tau + c_i)$ and $\xi_i \sim N(0,1)$. Expression (4.2) with $\theta_i = \theta$ is equivalent to the original $AR(1)$ process defined in (4.1) with the extra normality assumption in the errors; however, in our latent variable construction $\theta \in (0,1)$, since τ and c are both precision parameters, which means they are both positive, whereas in the original definition of the process, $\theta \in (-1,1)$. On the other hand, our latent variable construction with varying c_i (and thus θ_i) is a more general version that allows for different (positive) dependencies along time. If by starting from (4.1) and changing θ by θ_i it is not obvious that the process still had an invariant distribution, with our latent variable construction it is.

In this case, since the construction of the Markov process for $\{X_i\}$ is based on conditional independent normal distributions and marginal normal distributions for the conditioning variables, the joint (marginal) distribution for a finite collection n, say $\mathbf{X} = (X_i, \ldots, X_n)$, is multivariate normal, that is,

$$\mathbf{X} \sim N_n(\mu\mathbf{1}, \mathbf{C}),$$

where $\mathbf{1}$ is a vector of ones of dimension n and $\mathbf{C}^{-1}$ is the variance-covariance matrix with diagonal elements $1/\tau$ and off-diagonal elements $\rho_{i,i+s}/\tau$ with $\rho_{i,i+s} = \prod_{j=0}^{s-1} c_{i+j}/\tau(\tau + c_{i+j})$.

To have an idea of how these Markov processes behave, in Figures 4.3 and 4.4 we present five simulated paths for the beta process $\{X_i\}$ of Example 4.2 for $i = 1, \ldots, n$ and $n = 20$ in panel (a) of and the autocorrelation function $\rho(l) = \text{Corr}(X_1, X_{1+l})$ for $l = 1, 2, \ldots, 19$ in panel (b). In Figure 4.3 we took uniform marginal distributions, obtained by taking $a = b = 1$, but with different strengths of dependence. In row (a) we assumed an independent sequence, obtained with $c_i = 0$; in the rows (b) and (c) we defined Markov processes with parameters $c_i = 1$ and $c_i = 5$, respectively. In the independence case, the path quickly moves from one place to another in

4.2 Examples

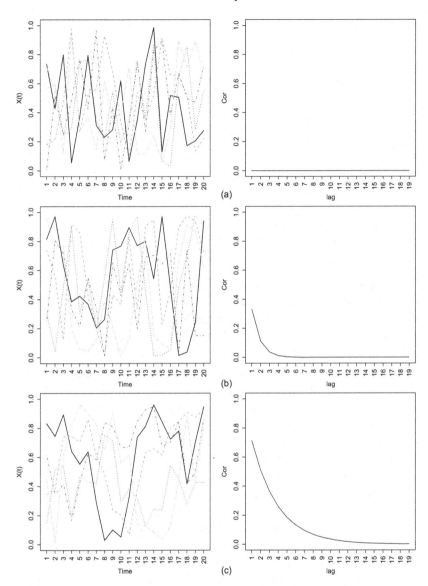

Figure 4.3 Simulations. Five paths (left) and autocorrelation function (right). $a = b = 1$, $c_i = 0$ (**a**), $c_i = 1$ (**b**) and $c_i = 5$ (**c**).

the space $[0, 1]$, and as we increase the dependence, the paths tend to move slower. In the right-hand-side panels we see that the dependence induced by our Markov construction is decreasing as a function of the lag l, starting at $1/3$ for $c_i = 1$ and $5/7 \approx 0.71$ for $c_i = 5$.

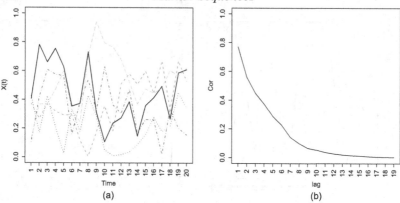

Figure 4.4 Simulations. Five paths **(a)** and autocorrelation function **(b)**. $a = 1$, $b = 2$, $c_i \sim \text{Po}(10)$.

In Figure 4.4 we took a right skewed beta as marginal distribution, obtained with $a = 1$ and $b = 2$ and with random association parameters c_i. In particular, we took $c_i \sim \text{Po}(10)$. The paths are more concentrated around $1/3$, which is the mean of the beta distribution, and the correlation does not decrease smoothly as it depends on the simulated values c_i.

In traditional time series analysis (e.g. Box et al. (2015)), the order of dependence p in an autoregressive process is determined via the sample partial autocorrelation function denoted as $PACF(l)$ for lags $l = 1, \ldots, n$. This is defined as the autocorrelation between X_i and X_{i+l}, where the linear dependence on $X_{i+1}, \ldots, X_{i+l-1}$ has been removed. A plot of $PACF(l)$ versus l for $l = 1, \ldots, L$, with L a lot lower than n, has a monotonic decay towards zero with $PACF(l) \neq 0$ for $l \leq p$ and $PACF(l) = 0$ for $l > p$.

Our Markov constructions are autoregressive processes or order $p = 1$, so for large n, a graph of the partial autocorrelation function $PACF(l)$ versus l should vanish for lags $l > 1$. For the simulated scenarios of Figures 4.3 and 4.4, we extended the sample size to $n = 300$ and computed the sample partial autocorrelation function $PACF(l)$ for $l = 1, \ldots, 20$. These are shown in Figure 4.5. Recall that scenarios considered vary the strength of dependence parameters c_i. For $c_i = 0$ (panel **(a)**) the PACF does not detect any order of dependence because data are independent. For $c_i = 1$ and $c_i = 5$ (panels **(b)** and **(c)**) the PACF properly detects the order of dependence $p = 1$ with $PACF(l)$ values outside the confidence intervals for $l = 1$ and practically zero for $l \geq 2$. When c_i is random and taken from a $\text{Po}(10)$, although the values of c_i are all different for each i, the PACF is able to properly detect the order of dependence $p = 1$. This is surprising

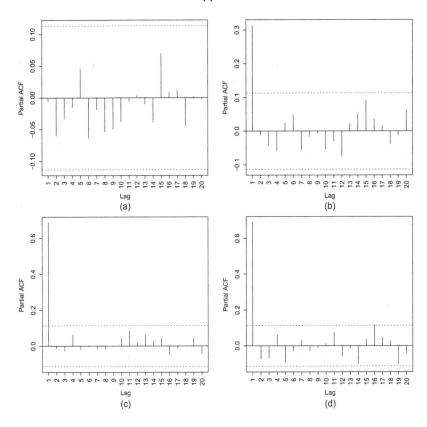

Figure 4.5 PACF of simulated paths of size $n = 300$. $a = b = 1$ with $c_i = 0$ (a), $c_i = 1$ (b), $c_i = 5$ (c) and $a = 1$, $b = 2$, $c_i \sim \text{Po}(10)$ (d).

because the sample PACF estimates a correlation that is fixed, but in our model it varies according to the different c_i.

4.3 Applications

There are several applications of these Markov sequences, but we present only one. This deals with survival analysis under a Bayesian nonparametric approach.

We start by recalling some survival analysis concepts. Let T be a nonnegative random variable with density function $f(t)$ and CDF $F(t)$. We define the survival function as $S(t) = P(T > t) = 1 - F(t)$. Let $h(t)$ be the hazard function which in discrete time is defined as $h(t) = P(T = t \mid T \geq t)$ and in continuous time as $h(t) = \lim_{\epsilon \to 0} P(t < T \leq t + \epsilon \mid T \geq t)$.

Let us assume that T is discrete with support $\{\tau_1, \tau_2, \ldots, \tau_J\}$, where J could be finite or infinite. In this context, let $h_j = h(\tau_j)$, the hazard function evaluated at the point mass τ_j. We can re-write the density and survival functions in terms of the hazard rates as

$$f(t) = h_j \prod_{k<j}(1 - h_k)I(t = \tau_j) \quad \text{and} \quad S(t) = \prod_{k \leq j}(1 - h_k)I(\tau_j \leq t < \tau_{j+1}).$$

We can treat $\mathbf{h} = \{h_j\}$ as the 'parameters' of the model and carry out a Bayesian inference.

Given a sample of possible right censored observations (t_i, δ_i), $i = 1, \ldots, n$, where δ_i is the censored indicator which takes the value of one if the observation t_i is exact and zero if it is right censored, the likelihood is

$$f(\mathbf{t} \mid \mathbf{h}) = \prod_{i=1}^{n} f(t_i)^{\delta_i} S(t_i)^{1-\delta_i} = \prod_{j=1} h_j^{n_j}(1 - h_j)^{m_j},$$

where $n_j = \sum_{i=1}^{n} I(t_i = \tau_j, \delta_i = 1)$ is the number of individuals who die at time τ_j and $m_j = \sum_{i=1}^{n} I(t_i \geq \tau_j)$ is the number of individuals at risk at time τ_j. Nieto-Barajas and Walker (2002) suggested using the Markov beta process of Example 4.2 as the prior distribution for $\{h_j\}$. This is obtained as the marginal distribution of the following joint:

$$f(\mathbf{h}, \mathbf{y}) = \text{Be}(h_1 \mid a, b) \prod_{j=1}^{J} \text{Bin}(y_j \mid c_j, h_j)\text{Be}(h_{j+1} \mid a + y_j, b + c_j - y_j).$$

We do not need to marginalise the latent variables $\mathbf{y}$; instead we treat $\mathbf{y}$ as latent parameters and characterise the posterior distribution $f(\mathbf{h}, \mathbf{y} \mid \mathbf{t})$ via their full conditional distributions and implement a Gibbs sampler (Smith and Roberts (1993)). The posterior conditional distributions are

$$f(h_j \mid \mathbf{t}, \text{rest}) = \text{Be}(a + y_{j-1} + y_j + n_j, b + c_{j-1} - y_{j-1} + c_j - y_j + m_j)$$

and

$$f(y_j \mid \mathbf{t}, \text{rest}) \propto \frac{\left\{\frac{h_j h_{j+1}}{(1-h_j)(1+h_{j+1})}\right\}^{y_j} I_{\{0,1,\ldots,c_j\}}(y_j)}{\Gamma(y_j + 1)\Gamma(c_j - y_j + 1)\Gamma(a + y_j)\Gamma(b + c_j - y_j)}$$

for $j = 1, \ldots, J$.

On the other hand, if T is continuous with support on the whole $\mathbb{R}^+$, our approach requires a finite partition of the support of the form $\mathbb{R}^+ = \cup_{j=1}^{J}(\tau_{j-1}, \tau_j]$, where $0 = \tau_0 < \tau_1 < \cdots < \tau_J = \infty$. We define a piecewise constant hazard function

$$h(t) = \sum_{j=1}^{J} h_j I_{(\tau_{j-1}, \tau_j]}(t)$$

and recover the density and survival functions with traditional expressions $f(t) = h(t)S(t)$ and $S(t) = \exp\{-\int_0^\infty h(s)ds\}$. Again, treating $\mathbf{h} = \{h_j\}$ as the parameters of the model, we carry out Bayesian inference.

Given a sample of possible right censored observations (t_i, δ_i), $i = 1, \ldots, n$, where δ_i is the censored indicator, the likelihood is

$$f(\mathbf{t} \mid \mathbf{h}) = \prod_{i=1}^{n} f(t_i)^{\delta_i} S(t_i)^{1-\delta_i} = \prod_{j=1} h_j^{n_j}(1 - h_j)^{m_j},$$

where $n_j = \sum_{i=1}^{n} I(\tau_{j-1} < t_i \leq \tau_j, \delta_i = 1)$ is the number of individuals who die at interval $(\tau_{j-1}, \tau_j]$ and $m_j = \sum_{i=1}^{n}(\tau_j - \tau_{j-1})I(t_i > \tau_j) + (t_i - \tau_{j-1})I(\tau_{j-1} < t_i \leq \tau_j)$ is the length time at risk for interval $(\tau_{j-1}, \tau_j]$. Nieto-Barajas and Walker (2002) suggested using the Markov gamma process of Example 4.4 as the prior distribution for $\{h_j\}$. This is obtained as the marginal distribution of the following joint:

$$f(\mathbf{h}, \mathbf{y}) = \text{Ga}(h_1 \mid a, b) \prod_{j=1}^{J} \text{Po}(y_j \mid c_j h_j) \text{Ga}(h_{j+1} \mid a + y_j, b + c_j).$$

As earlier, we do not need to marginalise the latent variables $\mathbf{y}$; we treat $\mathbf{y}$ as latent parameters and characterise the posterior distribution $f(\mathbf{h}, \mathbf{y} \mid \mathbf{t})$ via their full conditional distributions. These are

$$f(h_j \mid \mathbf{t}, \text{rest}) = \text{Ga}(a + y_{j-1} + y_j + n_j, b + c_{j-1} + c_j + m_j)$$

and

$$f(y_j \mid \mathbf{t}, \text{rest}) \propto \frac{\{c_j(b + c_j)h_j h_{j+1}\}^{y_j}}{\Gamma(y_j + 1)\Gamma(a + y_j)} I_{0,1,\ldots}(y_j)$$

for $j = 1, \ldots, J$.

The original contribution in Nieto-Barajas and Walker (2002) defines the Markov beta and gamma processes with general parameters a_j and b_j. This implies that the processes are still Markov, but do not have an invariant distribution. However, it is possible to allow for extra flexibility for modelling purposes.

Posterior inference with these two models is available in the R-package *BGPhazard* (Morones-Ishikawa et al. (2021)). We now present a specific data analysis.

6-MP Acute Leukemia Data Set

To illustrate the performance of the Markov beta and gamma processes in survival analysis, we consider the 6-MP data set (Freireich et al. (1963)) available in Table 2 of the Appendix. This consists of results from a clinical trial of 42 children with acute leukemia in complete or partial remission from their leukemia induced by a treatment with the drug prednisone. A remission status is achieved when most of the signs of the disease have disappeared from the bone marrow. The trial was conducted by matching pairs of patients by remission status (complete or partial) and randomising within the pair to either a 6-MP or a placebo maintenance therapy. Patients were followed until their leukemia relapsed (returned) or until the end of the study (in months).

Since there are two treatment groups, control and 6-MP, we analyse each of them separately. The control group only has exact observations and will be analysed assuming a discrete survival model with support defined by $\tau_j = j, j = 1, \ldots, J$ with $J = 23$. The prior distribution for the hazard rates $\{h_j\}$ is defined by a Markov beta process with $a = b = 0.0001$ and two different values for the dependence parameters $c_j \in \{2, 20\}$ to compare. The Gibbs sampler was run for 3,000 iterations, a burn-in of 300 and keeping one of every 2nd iteration to produce inference. The R command to implement this model is BeMRes and the complete list of commands is included in Table 4.1.

The objective is to estimate the survival curve for the control group and compare it with the Kaplan–Meier (KM) estimate (Kaplan and Meier (1958)), which is the most common nonparametric estimator in practice. On the other hand, we also want to show the effect of a small/large dependence parameter in the posterior inference. Estimators are reported in Figure 4.6. The solid light line corresponds to the KM estimator, together with its 95% confidence interval as light dotted lines. As we can see, the KM only jumps at times where there are observations. It continuously jumps from 1 to 5, remains constant from 5 to 8, where it jumps again, remaining constant until time 11 months and so on. At times 6, 7, 9–10, 13–14, 16 and 18–21 there were no observations, so there is no information to estimate the hazards h_j at those times.

What the Markov beta process prior does is to borrow strength from the neighbouring times and estimate the hazard rate everywhere, even at times where no data were observed. When $c_j = 2$, the dependence is low, but the survival function estimate jumps in all times of the support from 1 to 23. This is shown as the dark line in panel (**a**) of Figure 4.6, where the Bayesian

4.3 Applications

```
install.packages("BGPhazard")
library(BGPhazard)
data(gehan)
#Discrete
timesC <- gehan$time[gehan$treat == "control"]
deltaC <- gehan$cens[gehan$treat == "control"]
gehanC1 <- BeMRes(timesD,deltaD,c.r=rep(2,22),type.c=2,
                  iterations=3000,burn.in=300,thinning=2)
BePloth(gehanC1)
gehanC2 <- BeMRes(timesD,deltaD,c.r=rep(20,22),type.c=2,
                  iterations=3000,burn.in=300,thinning=2)
BePloth(gehanC2)
#Continuous
timesT <- gehan$time[gehan$treat == "6-MP"]
deltaT <- gehan$cens[gehan$treat == "6-MP"]
gehanT1 <- GaMRes(timesT,deltaT,type.t=1,K=5,
                  iterations=3000)
GaPloth(gehanT1)
```

Table 4.1 *R code for implementing survival analysis with beta and gamma Markov priors.*

nonparametric estimate is below the KM, but the interval estimates cross, which means that both estimators are practically the same.

When $c_j = 20$, the dependence is higher, producing a stronger exchange of information across times. The survival function estimate, shown in panel **(b)** of Figure 4.6, is practically overlapped with the KM, but in a smoother way and with jumps everywhere in the support. Additionally, interval estimates are a lot narrower than those of the KM.

Survival times from the treatment (6-MP) group have 9 exact observations (43%) and 12 right censored observations (57%). To analyse these data we assume a continuous time survival model, where the positive real line is partitioned into $J = 5$ intervals defined by $\tau_0 = 0$, $\tau_1 = 6$, $\tau_2 = 10$, $\tau_3 = 16$, $\tau_4 = 23$ and $\tau_5 = \infty$. These limits were obtained such that the number of exact observations in each interval for $j < J$ is approximately the same. This is achieved by the command `GaMRes` and the argument `time.t=1`. The Markov gamma process prior for piecewise constant hazard function is defined with $a = b = 0.0001$ and hierarchical $c_j \mid d \sim \text{Ga}(1, 1/d)$ with $d \sim \text{Ga}(0.1, 0.1)$. The Gibbs sampler had the same specifications as earlier and the complete list of commands is also included in Table 4.1.

Posterior estimates are shown in Figure 4.7. In panel **(a)** we include estimates of the hazard function $h(t)$. It is clearly piecewise and has a lot

Markov Sequences

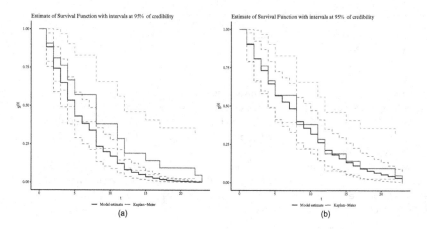

Figure 4.6 6-MP drug leukemia data set. Survival function estimates. KM (light lines) and Markov model (dark lines). Point (solid line) and 95% credible intervals (dotted lines). $c_j = 2$ (panel **(a)**), $c_j = 20$ (panel **(b)**).

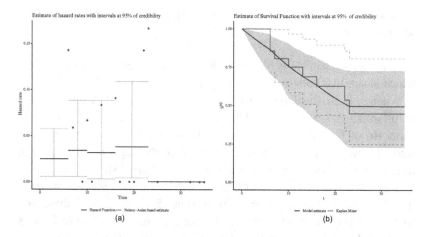

Figure 4.7 6-MP drug leukemia data set. Hazard rate (panel **(a)**) and survival function estimates (panel **(b)**). **(a)** Nelson–Aalen point estimates (dots), Markov model point estimates (dark lines) and 95% CI (light lines). **(b)** KM point estimate (light line) and 95% CI (dotted line), Markov model point estimate (dark solid) and 95% CI (shadows).

of uncertainty with very wide credible intervals. This is due to the small sample size of exact observations in each interval, three data points in the first interval, two in the middle ones and zero in the last one. For comparison, the light dots correspond to the Nelson–Aalen nonparametric estimators of

4.3 Applications

the hazard rates, which are totally different to our Bayesian estimators. However, survival curve estimates look similar. The KM estimate is a stepwise function because it assumes an underlying discrete survival model, whereas the estimate with our model is a smooth continuous function that follows the path of the KM because we assume a continuous underlying survival model.

A different application of the Markov constructions was considered by Mendoza and Nieto-Barajas (2006) in a solvency analysis for insurance companies.

5

General Dependent Sequences

5.1 First Attempts

Markov processes are also useful for modelling time series data; however, since the dependence is of order one, sometimes this could be restrictive. We formulated for ourselves the following question: is it possible to define a dependence process such that the order of dependence is larger than one and maintain stationarity with a desired invariant distribution? In a graphical model the question is, can we place an arrow from X_1 to X_2 and X_3 and from X_2 to X_3 and X_4, and so on? This is depicted in Figure 5.1.

One solution is to consider an autoregressive process of order 2, $AR(2)$. This is defined as

$$X_t = \theta_1 X_{t-1} + \theta_2 X_{t-2} + Z_t,$$

where the Z_t are random variables such that $E(Z_t) = 0$, $Var(Z_t) = \sigma^2$ and $Cov(Z_t, Z_s) = 0$ for all $t \neq s$. If the index space is unbounded, say $t \in \mathbb{Z}$, it is possible to impose certain conditions on the parameters (θ_1, θ_2) such that the process $\{X_t\}$ is weakly stationary (see Box et al. (2015)). However, if the index set is bounded, say $t = 1, \ldots, n$, it is not possible to define stationarity in the $AR(2)$ process.

Another possibility is to consider the latent variables approach introduced in Chapter 4. Instead of linking directly X_1 to X_2 and X_3, we link the latent variable Y_1 with X_2 and X_3. This is depicted graphically in Figure 5.2.

To illustrate what would happen, let us consider the normal conjugate family.

Example 5.1 Let X_1, Y_1, X_2, Y_2 be defined as in Example 4.6, that is, $X_1 \sim N(\mu, \tau)$, $Y_1 \mid X_1 \sim N(x_1, c_1)$, $X_2 \mid Y_1 \sim N((\tau\mu + c_1 y_1)/(\tau + c_1), \tau + c_1)$ and $Y_2 \mid X_2 \sim N(x_2, c_2)$. This implies that marginally Y_1 and Y_2 are $N(\mu, \tau c_i/(\tau + c_i))$ for $i = 1, 2$ and X_1 and X_2 are $N(\mu, \tau)$. Now, for the rest of the variables we could define $Y_i \mid X_i \sim N(x_i, c_i)$ and

5.1 First Attempts

Figure 5.1 Graphical representation of an order 2 dependent process.

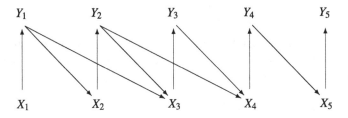

Figure 5.2 Graphical representation of an order 2 dependent process via latent variables.

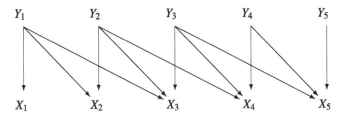

Figure 5.3 Graphical representation of a moving average process of order $q = 2$.

$$X_i \mid Y_{i-1}, Y_{i-2} \sim N\left(\frac{\tau\mu + c_{i-1}y_{i-1} + c_{i-2}y_{i-2}}{\tau + c_{i-1} + c_{i-2}}, \tau + c_{i-1} + c_{i-2}\right)$$

for $i = 3, 4, \ldots$, which makes sense since the latent variables act as the likelihood and the sufficient statistic is the sum $Y_{i-1} + Y_{i-2}$. If we obtain the marginal distribution for X_3, it will not be $N(\mu, \tau)$ because Y_1 and Y_2 are conditionally independent, given X_2 and X_3, and the process is not reversible anymore.

A third possible solution would be to reverse the direction of some arrows in Figure 5.2 so that the latent variables Y_i do not have incoming arrows and the X_i do not have outgoing arrows. This is depicted in Figure 5.3. This new construction is not of an autoregressive nature, but becomes similar to a moving average process such as the one presented in Example 1.9 with $q = 2$.

Again, let us consider the normal conjugate family to illustrate.

72 General Dependent Sequences

Example 5.2 Let $\mathbf{Y} = (Y_1, Y_2, \ldots)$ be independent random variables such that $Y_i \sim N(\mu, c_i)$, for $i = 1, 2, \ldots$. One way of defining the X_i is considering the form of the posterior distribution for the location parameter in a conjugate normal model. That is,

$$X_i \mid \mathbf{Y} \sim N\left(\frac{\tau\mu + \sum_{j=0}^{q} c_{i-j} y_{i-j}}{\tau + \sum_{j=0}^{q} c_{i-j}}, \tau + \sum_{j=0}^{q} c_{i-j}\right)$$

for $i = 1, 2, \ldots$ and define $Y_0 = 0, Y_{-1} = 0, \ldots, Y_{1-p} = 0$ with probability one (w.p.1) and $c_0 = \cdots = c_{1-p} = 0$. To see why this definition resembles a moving average process, let us re-write it is as follows:

$$X_i = \alpha_i + \sum_{j=0}^{q} \theta_{i-j} Y_{i-j} + \epsilon_i,$$

where $\alpha_i = \tau\mu/(\tau + \sum_{j=0}^{q} c_{i-j})$, $\theta_i = c_i/(\tau + \sum_{j=0}^{q} c_{i-j})$ and $\epsilon_i \sim N(0, \tau + \sum_{j=0}^{q} c_{i-j})$ for $i = 1, 2, \ldots$. Moreover, we can re-write the previous expression in terms of standard normal random variables, that is,

$$X_i = \alpha_i + \sum_{j=1}^{q} \vartheta_{i-j} Z_{i-j} + \beta_i \varepsilon_i,$$

where $\vartheta_i = \theta_i c_i^{-1/2}$, $\beta_i = (\tau + \sum_{j=0}^{q} c_{i-j})^{-1/2}$ with $Z_i \sim N(0, 1)$ and $\varepsilon_i \sim N(0, 1)$ independent for $i = 1, 2, \ldots$. It must now be clear why the definition of X_i is similar to a moving average process of order q plus innovation such as the one presented in Example 1.9. Now, let's compute the first two moments of X_i to see if we can achieve stationarity or an invariant distribution. Using iterative expectations, we get

$$E(X_i) = \frac{\tau\mu + \sum_{j=1}^{q} c_{i-j} E(Y_{i-j})}{\tau + \sum_{j=0}^{q} c_{i-j}} = \mu$$

and the variance becomes

$$\mathrm{Var}(X_i) = E\left(\frac{1}{\tau + \sum_{j=0}^{q} c_{i-j}}\right) + \mathrm{Var}\left(\frac{\tau\mu + \sum_{j=0}^{q} c_{i-j} Y_{i-j}}{\tau + \sum_{j=0}^{q} c_{i-j}}\right)$$

$$= \frac{1}{\tau + \sum_{j=0}^{q} c_{i-j}} + \frac{\sum_{j=0}^{q} c_{i-j}}{\left(\tau + \sum_{j=0}^{q} c_{i-j}\right)^2}$$

$$= \frac{\tau + 2\sum_{j=0}^{q} c_{i-j}}{\left(\tau + \sum_{j=0}^{q} c_{i-j}\right)^2}.$$

5.1 First Attempts

Using iterative covariance formulae and conditional independence, the covariance between X_i and X_{i+k} is

$$\begin{aligned}\text{Cov}(X_i, X_{i+k}) &= \text{Cov}\{E(X_i \mid \mathbf{Y}), E(X_{i+k} \mid \mathbf{Y})\} \\ &= \frac{\text{Cov}\left(\sum_{j=1}^{q} c_{i-j} E(Y_{i-j}), \sum_{j=1}^{q} c_{i+k-j} E(Y_{i+k-j})\right)}{(\tau + \sum_{j=0}^{q} c_{i-j})(\tau + \sum_{j=0}^{q} c_{i+k-j})} \\ &= \frac{\text{Var}\left(\sum_{j=0}^{q-k} c_{i-j} Y_{i-j}\right)}{(\tau + \sum_{j=0}^{q} c_{i-j})(\tau + \sum_{j=0}^{q} c_{i+k-j})} \\ &= \frac{\sum_{j=0}^{q-k} c_{i-j}}{(\tau + \sum_{j=0}^{q} c_{i-j})(\tau + \sum_{j=0}^{q} c_{i+k-j})}\end{aligned}$$

for $k \leq q$, and zero otherwise. Finally, the correlation between X_i and X_{i+k}, for $k \leq q$, becomes

$$\text{Corr}(X_i, X_{i+k}) = \frac{\sum_{j=0}^{q-k} c_{i-j}}{\sqrt{(\tau + 2\sum_{j=0}^{q} c_{i-j})(\tau + 2\sum_{j=0}^{q} c_{i+k-j})}}.$$

Now, since X_i is a linear combination of independent random variables, the marginal distribution of X_i will be normal with mean μ, but with a variance that depends on i and a correlation between (X_i, X_{i+k}) that vanishes for $k > q$. Although we have marginal normality, we do not have an invariant distribution and therefore have no stationarity.

A different example where we do have an invariant distribution in a moving average–type process, such as the one in Figure 5.3, is for the Poisson case and was proposed by Nieto-Barajas (2022b).

Example 5.3 Let $\mathbf{Y} = (Y_1, Y_2, \ldots)$ be independent random variables such that $Y_i \sim \text{Po}(\mu \alpha_i)$ for $i = 1, 2, \ldots$. We define the random variables of interest, X_i, using ideas from the quasi-conjugate Poisson binomial model as in Examples 1.3 and 4.5, that is,

$$X_i - \sum_{j=0}^{q} y_{i-j} \mid \mathbf{Y} \sim \text{Po}\left(\mu\left(1 - \sum_{j=0}^{q} \alpha_{i-j}\right)\right)$$

for $i = 1, 2, \ldots$, and define $Y_0, \ldots, Y_{1-q}$ equal to zero w.p.1 and $c_0 = \cdots = c_{1-q} = 0$. Again, this definition of the X_i resembles a moving average construction of order q. To obtain the marginal distribution of X_i we first note that the sum of independent Poisson random variables is also Poisson with parameter the sum of the individual parameters, that is, $\sum_{j=0}^{q} Y_{i-j} \sim$ Po $\left(\mu \sum_{j=0}^{q} \alpha_{i-j}\right)$. Now, recalling the result that

if $X \sim \text{Po}(\mu)$ and $Y \mid X \sim \text{Bin}(x, \alpha)$

$\iff Y \sim \text{Po}(\mu\alpha)$ and $X - y \mid Y \sim \text{Po}(\mu(1 - \alpha))$,

therefore $X_i \sim \text{Po}(\mu)$ for $i = 1, 2, \ldots$. Moreover, using iterative covariance formulae, the covariance between X_i and X_{i+k} is

$$\text{Cov}(X_i, X_{i+k}) = \text{Cov}\{E(X_i \mid \mathbf{Y}), E(X_{i+k} \mid \mathbf{Y})\}$$

$$= \text{Cov}\left\{\mu\left(1 - \sum_{j=0}^{q} \alpha_{i-j}\right) + \sum_{j=0}^{q} Y_{i-j},\right.$$

$$\left.\mu\left(1 - \sum_{j=0}^{q} \alpha_{i+k-j}\right) + \sum_{j=0}^{q} Y_{i+k-j}\right\}$$

$$= \text{Var}\left(\sum_{j=0}^{q-k} Y_{i-j}\right) = \mu \sum_{j=0}^{q-k} \alpha_{i-j}$$

for $k \leq q$ and zero otherwise. Finally, the correlation for $k \leq q$ is

$$\text{Corr}(X_i, X_{i+k}) = \sum_{j=0}^{q-k} \alpha_{i-j}.$$

Therefore, $\mathbf{X} = \{X_i\}$ is a $MA(q)$-type process with $\text{Po}(\mu)$ invariant marginal distribution. Differing from the $MA(q)$ process defined in Example 1.9, where there were $q + 1$ coefficients, here we have as many coefficients as elements in the sequence $\mathbf{X}$.

5.2 Main Results

Instead of assuming independence in the Y_i, Jara et al. (2013) suggested taking a dependence sequence, as in Chapter 3, defined conditionally independent given a common latent variable, say Z, that acts as an anchor for all latent variables Y_i, but with the possibility of the Y_i having each a different parameter. The desired sequence X_i is then defined via conditional independence given as many Y_i as desired to define an order q dependent sequence. Graphically a dependence process of order $q = 2$ is shown in Figure 5.4.

By considering appropriate definitions of the conditional distributions in a three-level hierarchical model, we can achieve an invariant distribution. This is given in the following proposition.

5.2 Main Results

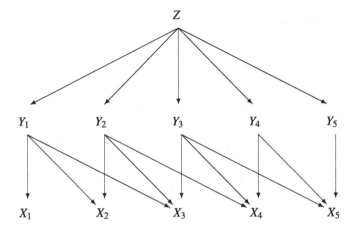

Figure 5.4 Graphical representation of dependence model of order $q = 2$.

Proposition 5.4 *Let X be a random variable with known distribution $f_X(x)$. Let Y be a latent variable with arbitrary conditional distribution $f_{Y|X}(y \mid x)$. Let $\mathbf{Y} = (Y_1, \ldots, Y_n)$ be a collection of size n of conditionally independent latent variables such that $Y_i \mid X \sim f_{Y|X}(y_i \mid x)$, and let $f_{X|\mathbf{Y}}(x \mid \mathbf{y})$ be the corresponding conditional distribution for X given $\mathbf{Y}$ obtained via Bayes's theorem, that is,*

$$f_{X|\mathbf{Y}}(x \mid \mathbf{y}) = \frac{f_{\mathbf{Y}|X}(\mathbf{y} \mid x) f_X(x)}{f_\mathbf{Y}(\mathbf{y})},$$

where $f_{\mathbf{Y}|X}(\mathbf{y} \mid x) = \prod_{i=1}^n f_{Y|X}(y_i \mid x)$ and with $f_\mathbf{Y}(\mathbf{y}) = \int f_{\mathbf{Y}|X}(\mathbf{y} \mid x) f_X(x) \mathrm{d}x$ or $f_\mathbf{Y}(\mathbf{y}) = \sum_x f_{\mathbf{Y}|X}(\mathbf{y} \mid x) f_X(x)$ according to whether X is continuous or discrete, respectively. Let Z and $\mathbf{Y} = \{Y_i\}$, two sets of latent variables, and let $\mathbf{X} = \{X_i\}$, the sequence of interest defined via a three-level hierarchical model of the form

$$Z \sim f_X(z),$$

$$Y_i \mid Z \stackrel{ind}{\sim} f_{Y|X}(y_i \mid z)$$

conditionally independent, given Z for $i = 1, 2 \ldots$, and

$$X_i \mid \mathbf{Y}_i^* \stackrel{ind}{\sim} f_{X|Y_i^*}(x_i \mid \mathbf{y}_i^*),$$

conditionally independent, where $\mathbf{Y}_i^*$ is a subset of $\{Y_j\}$ of size $n_i \neq 0$. Then the marginal distribution of X_i is invariant and coincides with $f_X(x)$, regardless of the subsets chosen.

76   *General Dependent Sequences*

Proof Let us define $\partial_i = \{j : Y_j \in \mathbf{Y}_i^*\}$ the set of indexes of the Y_j variables that belong to subset $\mathbf{Y}_i^*$. Let us assume that the latent variables Y_i and Z are continuous, then the marginal distribution for X_i is obtained by integrating the rest of the variables in the joint for $(X_i, \mathbf{Y}_i^*, Z)$, that is,

$$f_{X_i}(x_i) = \int \int f_{X_i, \mathbf{Y}_i^*, Z}(x_i, \mathbf{y}_i^*, z) \, d\mathbf{y}_i^* \, dz$$

$$= \int \int f_{X_i | \mathbf{Y}_i^*}(x_i \mid \mathbf{y}_i^*) f_{\mathbf{Y}_i^* | Z}(\mathbf{y}_i^* \mid z) f_Z(z) \, d\mathbf{y}_i^* \, dz$$

$$= \int \int f_{X_i | \mathbf{Y}_i^*}(x_i \mid \mathbf{y}_i^*) \prod_{j \in \partial_i} f_{Y_j | X}(y_j \mid z) f_X(z) \, d\mathbf{y}_i^* \, dz.$$

Using Bayes's theorem to express $f_{X_i | \mathbf{Y}_i^*}(x_i \mid \mathbf{y}_i^*)$ and re-grouping some terms, we have

$$f_{X_i}(x_i) = \int \frac{f_{\mathbf{Y}_i^* | X}(\mathbf{y}_i^* \mid x_i) f_X(x_i)}{f_{\mathbf{Y}_i^*}(\mathbf{y}_i^*)} \left\{ \int \prod_{j \in \partial_i} f_{Y_j | X}(y_j \mid z) f_X(z) \, dz \right\} d\mathbf{y}_i^*.$$

We note that the term in curly parentheses { } coincides with $f_{\mathbf{Y}_i^*}(\mathbf{y}_i^*)$; then

$$f_{X_i}(x_i) = \int f_{\mathbf{Y}_i^* | X}(\mathbf{y}_i^* \mid x_i) f_X(x_i) \, d\mathbf{y}_i^*.$$

Since the integral is with respect to $\mathbf{y}_i^*$, we can factorise $f_X(x_i)$ to obtain

$$f_{X_i}(x_i) = f_X(x_i) \int f_{\mathbf{Y}_i^* | X}(\mathbf{y}_i^* \mid x_i) \, d\mathbf{y}_i^* = f_X(x_i)$$

because $f_{\mathbf{Y}_i^* | X}(\mathbf{y}_i^* \mid x_i)$ is a density that integrates up to one, which completes the proof. □

Proposition 5.4 is a powerful tool that allows us to construct dependence sequences $\{X_i\}$ with a desired invariant distribution $f_X(x)$ via a three-level hierarchical model and an appropriate use of Bayes's theorem. Again, the first level is associated with the prior, the second to the likelihood and the third to the posterior. The dependence across X_i is determined by the subset of latent variables $\mathbf{Y}_i^*$ chosen for each i. For instance, to define an order q process for time series analysis, the subsets can be defined as $\mathbf{Y}_i^* = \{Y_i, Y_{i-1}, \ldots, Y_{i-q}\}$, where in particular, for $q = 2$, we obtain the model represented in Figure 5.4. However, the dependence does not need to be as in a moving average process; it could be asymmetric, seasonal, periodic, spatial or spatio-temporal, by an appropriate choice of sets $\mathbf{Y}_i^*$.

One way of characterising the dependence across X_i, induced by a particular choice of subsets $\mathbf{Y}_i^*$, is through the correlation; however, this

5.2 Main Results

measure depends on the particular marginal distribution chosen, $f_X(x)$. Nieto-Barajas and Gutiérrez-Peña (2022) characterised the correlation when $f_X(x)$ is the conjugate family for the exponential family with quadratic variance function. To state their results, we introduce some notation.

Let $U_1, U_2, \ldots, U_n$ be a sample from an exponential family with density

$$f(u \mid \theta(\mu)) = a(u) \exp\{\theta(\mu) s(u) - M(\theta(\mu))\},$$

where $a(\cdot)$ is a non-negative function, θ is the canonical parameter written in terms of the mean parameterisation $\mu \in \mathcal{M}$ with $\mathcal{M}$ the mean parameter space, $S(U)$ is the canonical statistic and $M(\theta)$ is the cumulant transform such that

$$\mu = E(S \mid \theta) = \frac{\partial}{\partial \theta} M(\theta) \quad \text{and} \quad V(\mu) = \text{Var}(S \mid \theta) = \frac{\partial^2}{\partial \theta^2} M(\theta),$$

where $V(\mu)$ is called variance function written in terms of the mean μ. Morris (1982) characterised six families whose variance function is quadratic, that is,

$$V(\mu) = v_0 + v_1 \mu + v_2 \mu^2. \tag{5.1}$$

These families will be described in detail in Section 5.3.

Let $S_n = \sum_{i=1}^{n} S(U_i)$ be the sufficient statistic for the preceding exponential family density, then the likelihood becomes

$$f(s_n \mid \theta, n) = b(s_n, n) \exp\{\theta(\mu) s_n - n M(\theta(\mu))\}. \tag{5.2}$$

The conjugate family for the parameter $\mu \in \mathcal{M}$ is

$$f(\mu \mid s_0, n_0) = h(s_0, n_0) \exp\{\theta(\mu) s_0 - n_0 M(\theta(\mu))\} |J_\theta(\mu)|, \tag{5.3}$$

with parameters s_0 and n_0, where $|J_\theta(\cdot)|$ is the Jacobian of the transformation $\theta(\cdot)$. Moments of this prior have general expressions if the variance function is quadratic. In particular, Diaconis and Ylvisaker (1979) and Morris (1983) showed that

$$E(\mu \mid s_0, n_0) = s_0/n_0 \quad \text{and} \quad \text{Var}(\mu \mid s_0, n_0) = V(s_0/n_0)/(n_0 - v_2),$$

respectively, where $V(\cdot)$ is given in (5.1) and v_2 is the coefficient of the quadratic term. The posterior distribution for μ has exactly the same form as the prior but with updated parameters $s^* = s_0 + s_n$ and $n^* = n_0 + n$.

We are now in a position to state the result.

Proposition 5.5 *Let Z and $\{Y_i\}$ be two sets of latent variables and let $\{X_i\}$ be the sequence of interest defined via a three-level hierarchical model of the form*

$$Z \sim f(z \mid s_0, c_0) = h(s_0, c_0) \exp\{\theta(z)s_0 - c_0 M(\theta(z))\} |J_\theta(z)|,$$

$$Y_i \mid Z \overset{ind}{\sim} f(y_i \mid \theta(z), c_i) = b(y_i, c_i) \exp\{\theta(z)y_i - c_i M(\theta(z))\},$$

conditionally independent, given Z for $i = 1, 2, \ldots$, and

$$X_i \mid \mathbf{Y}_i^* \overset{ind}{\sim} f(x_i \mid s_i^*, c_i^*) = h(s_i^*, c_i^*) \exp\{\theta(x_i)s_i^* - c_i^* M(\theta(x_i))\} |J_\theta(x_i)|,$$

conditionally independent, where

$$s_i^* = s_0 + \sum_{j \in \partial_i} y_j \quad \text{and} \quad c_i^* = c_0 + \sum_{j \in \partial_i} c_j,$$

where $\partial_i = \{j : Y_j \in \mathbf{Y}_i^*\}$ and $\mathbf{Y}_i^*$ is a subset of $\{Y_j\}$ for $i = 1, 2, \ldots$. Then the marginal distribution of X_i is invariant and coincides with $f(x \mid s_0, c_0)$ given in (5.3). Moreover, if the variance function is quadratic, the correlation between X_i and X_k is given by

$$\text{Corr}(X_i, X_k) = \frac{c_0 \left(\sum_{j \in \partial_i \cap \partial_k} c_j \right) + \left(\sum_{j \in \partial_i} c_j \right) \left(\sum_{j \in \partial_k} c_j \right)}{\left(c_0 + \sum_{j \in \partial_i} c_j \right) \left(c_0 + \sum_{j \in \partial_k} c_j \right)}.$$

Proof The invariance of the marginal distribution for X_i is obtained by Proposition 5.4. For the correlation we first note that $E(Y_i \mid Z) = c_i Z$, $\text{Var}(Y_i \mid Z) = c_i V(Z)$ with $V(\cdot)$ given in (5.1), $E(Z) = s_0/c_0$ and $\text{Var}(Z) = V(s_0/c_0)/(c_0 - v_2)$. Moreover, $E(X_i \mid \mathbf{Y}_i^*)$ and $\text{Var}(X_i \mid \mathbf{Y}_i^*)$ have the same form as the moments for Z but with (s_i^*, c_i^*) instead of (s_0, c_0). We first compute the covariance between X_i and X_k using the iterative formulae and obtain

$$\text{Cov}(X_i, X_k) = E\{\text{Cov}(X_i, X_k \mid \mathbf{Y})\} + \text{Cov}\{E(X_i \mid \mathbf{Y}), E(X_k \mid \mathbf{Y})\}.$$

The first term is zero due to conditional independence, so

$$\text{Cov}(X_i, X_k) = \text{Cov}\left(\frac{S_i^*}{c_i^*}, \frac{S_k^*}{c_k^*} \right) = \frac{1}{c_i^* c_k^*} \text{Cov}(S_i^*, S_k^*).$$

Using the iterative covariance formulae for a second time, we get

$$\text{Cov}(X_i, X_k) = \frac{1}{c_i^* c_k^*} \left[E\{\text{Cov}(S_i^*, S_k^* \mid Z)\} + \text{Cov}\{E(S_i^* \mid Z), E(S_k^* \mid Z)\} \right].$$

We split ∂_i and ∂_k into two disjoint sets, $\partial_i = (\partial_i \cap \partial_k) \cup (\partial_i \cap \partial_k^c)$ and $\partial_k = (\partial_k \cap \partial_i) \cup (\partial_k \cap \partial_i^c)$, where $(\partial_i \cap \partial_k) \cap (\partial_i \cap \partial_k^c) \cap (\partial_k \cap \partial_i^c) = \emptyset$. We can re-write the conditional covariance as

$$\text{Cov}(S_i^*, S_k^* \mid Z) = \text{Cov}(A, A \mid Z) + \text{Cov}(A, C \mid Z)$$
$$+ \text{Cov}(B, A \mid Z) + \text{Cov}(B, C \mid Z),$$

where

$$A = \sum_{j \in \partial_i \cap \partial_k} Y_j, \quad B = \sum_{j \in \partial_i \cap \partial_k^c} Y_j, \quad C = \sum_{j \in \partial_k \cap \partial_i^c} Y_j.$$

Since A, B and C do not share any random variable, their conditional covariances are zero, therefore

$$\text{Cov}(X_i, X_k) = \frac{1}{c_i^* c_k^*} \left[\text{E}\{\text{Var}(A \mid Z)\} + \text{Cov}\left(\sum_{j \in \partial_i} c_j Z, \sum_{j \in \partial_k} c_j Z \right) \right].$$

Given that $\text{Var}(A \mid Z) = \sum_{j \in \partial_i \cap \partial_k} \text{Var}(Y_j \mid Z) = \sum_{j \in \partial_i \cap \partial_k} c_j V(Z)$, then

$$\text{Cov}(X_i, X_k) = \frac{1}{c_i^* c_k^*} \left[\sum_{j \in \partial_i \cap \partial_k} c_j \text{E}\{V(Z)\} + \left(\sum_{j \in \partial_i} c_j \right) \left(\sum_{j \in \partial_k} c_j \right) \text{Var}(Z) \right].$$

Working on the expected value of the quadratic variance function, we get

$$\text{E}\{V(Z)\} = v_0 + v_1 \text{E}(Z) + v_2 \text{E}(Z^2) = v_0 + v_1 \frac{s_0}{c_0} + v_2 \left\{ \frac{V(s_0/c_0)}{n_0 - v_2} + \frac{s_0^2}{c_0^2} \right\}$$

$$= V(s_0/c_0) + \frac{v_2}{c_0 - v_2} V(s_0/c_0) = V(s_0/c_0) \frac{c_0}{c_0 - v_2} = c_0 \text{Var}(Z).$$

Going back to the covariance, we have

$$\text{Cov}(X_i, X_k) = \frac{1}{c_i^* c_k^*} \left\{ c_0 \left(\sum_{j \in \partial_i \cap \partial_k} c_j \right) + \left(\sum_{j \in \partial_i} c_j \right) \left(\sum_{j \in \partial_k} c_j \right) \right\} \text{Var}(Z).$$

Finally, recalling that $\text{Var}(X_i) = \text{Var}(X_k) = \text{Var}(Z)$, we get the result. □

5.3 Particular Exponential Family Cases

The exponential family with quadratic variance function includes six cases, known by the name of the conjugate family and the likelihood. Three of them have discrete likelihoods and the other three are continuous. They are gamma & Poisson, beta & binomial, inverse beta & negative binomial, inverse gamma & gamma, normal & normal (with known precision) and generalised scaled student & generalised hyperbolic secant distribution. Note that the conjugate family is for the mean parameter μ of the likelihood.

Family	Conjugate	Likelihood	$V(\mu)$	$\mathcal{M}$
(i) Gamma & Poisson	$\text{Ga}(s_0, c_0)$	$\text{Po}(\mu)$	μ	$\mathbb{R}^+$
(ii) Beta & Binomial	$\text{Be}(s_0, c_0 - s_0)$	$\text{Bin}(1, \mu)$	$\mu - \mu^2$	$(0, 1)$
(iii) Inv.Beta & Neg.Bin.	$\text{Ibe}(s_0, c_0 + 1)$	$\text{NB}\left(1, \frac{1}{1+\mu}\right)$	$\mu + \mu^2$	$\mathbb{R}^+$
(iv) Inv.Gam & Gamma	$\text{Iga}(c_0 + 1, s_0)$	$\text{Ga}\left(1, \frac{1}{\mu}\right)$	μ^2	$\mathbb{R}^+$
(v) Normal & Normal	$\text{N}\left(\frac{s_0}{c_0}, c_0\right)$	$\text{N}(\mu, 1)$	1	$\mathbb{R}$
(vi) G.S.St & G.Hyp.Sec.	$\text{GSSt}\left(\frac{s_0}{c_0}, c_0\right)$	$\text{GHS}(\mu, 1)$	$1 + \mu^2$	$\mathbb{R}$

Table 5.1 *The members of the natural exponential family with quadratic variance function.*

Table 5.1 presents a summary of these models together with the variance function (5.1) and the mean parameter space or domain $\mathcal{M}$.

Exponential family members (5.2) and their respective conjugate families (5.3) are characterised by functions $b(\cdot, \cdot)$, $h(\cdot, \cdot)$, $\theta(\cdot)$, $M(\cdot)$ and $J_\theta(\cdot)$. The following presents these explicit expressions for the six exponential family members with quadratic variance function. The names correspond to the conjugate family that determines the marginal invariant distribution in Proposition 5.5.

(i). *Gamma*: A dependence model with gamma invariant distribution uses latent variables whose conditional distribution is Poisson. This is characterised by functions $b(y_i, c_i) = c_i^{y_i}/y_i!$, $h(s_0, c_0) = c_0^{s_0}/\Gamma(s_0)$, $\theta(z) = \log(z)$, $M(\theta(z)) = z$ and $J_\theta(z) = 1/z$. In summary,

$$Z \sim \text{Ga}(s_0, c_0),$$
$$Y_i \mid Z \sim \text{Po}(c_i z),$$
$$X_i \mid \mathbf{Y}_i^* \sim \text{Ga}(s_i^*, c_i^*).$$

(ii). *Beta*: A dependence model with beta invariant distribution uses latent variables whose conditional distribution is binomial. This is characterised by functions $b(y_i, c_i) = \binom{c_i}{y_i}$, $h(s_0, c_0) = \Gamma(c_0)/\{\Gamma(s_0)\Gamma(c_0 - s_0)\}$, $\theta(z) = \log\{z/(1 - z)\}$, $M(\theta(z)) = -\log(1 - z)$ and $J_\theta(z) = 1/\{z(1 - z)\}$. In summary,

$$Z \sim \text{Be}(s_0, c_0 - s_0),$$
$$Y_i \mid Z \sim \text{Bin}(c_i, z),$$
$$X_j \mid \mathbf{Y}_i^* \sim \text{Be}(s_i^*, c_i^* - s_i^*).$$

5.3 Particular Exponential Family Cases

(iii). *Inverse Beta*: A dependence model with inverse beta invariant distribution uses latent variables whose conditional distribution is negative binomial. This is characterised by functions $b(y_i, c_i) = \binom{c_i + y_i - 1}{y_i}$, $h(s_0, c_0) = \Gamma(s_0 + c_0 + 1)/\{\Gamma(s_0)\Gamma(c_0 + 1)\}$, $\theta(z) = \log\{z/(z+1)\}$, $M(\theta(z)) = \log(z+1)$ and $J_\theta(z) = 1/\{z(z+1)\}$. In summary,

$$Z \sim \text{Ibe}(s_0, c_0 + 1),$$
$$Y_i \mid Z \sim \text{NB}(c_i, 1/(z+1)),$$
$$X_i \mid \mathbf{Y}_i^* \sim \text{Ibe}(s_i^*, c_i^* + 1).$$

(iv). *Inverse Gamma*: A dependent model with inverse gamma invariant distribution uses latent variables whose conditional distribution is gamma. This is characterised by functions $b(y_i, c_i) = s_i^{c_i - 1}/\Gamma(c_i)$, $h(s_0, c_0) = s_0^{c_0}/\Gamma(c_0)$, $\theta(z) = -1/z$, $M(\theta(z)) = \log(z)$ and $J_\theta(z) = 1/z$. In summary,

$$Z \sim \text{Iga}(c_0 + 1, s_0),$$
$$Y_i \mid Z \sim \text{Ga}(c_i, 1/z),$$
$$X_i \mid \mathbf{Y}_i^* \sim \text{Iga}(c_i^* + 1, s_i^*).$$

(v). *Normal*: A dependent model with normal invariant distribution uses latent variables whose conditional distribution is again normal. This is characterised by functions $b(y_i, c_i) = (2\pi c_i)^{-1/2} \exp\{-y_i^2/(2c_i)\}$, $h(s_0, c_0) = (2\pi/c_0)^{-1/2} \exp\{-s_0^2/(2c_0)\}$, $\theta(z) = z$, $M(\theta(z)) = z^2/2$ and $J_\theta(z) = 1$. In summary,

$$Z \sim \text{N}(s_0/c_0, c_0),$$
$$Y_i \mid Z \sim \text{N}(c_i z, 1/c_i),$$
$$X_i \mid \mathbf{Y}_i^* \sim \text{N}(s_i^*/c_i^*, c_i^*).$$

(vi). *Generalised Scaled Student*: A dependent model with generalised scaled student invariant distribution uses latent variables whose conditional distribution is generalised hyperbolic secant. This is characterised by functions $b(y_i, c_i) = \{2^{c_i - 2}/\Gamma(c_i)\} \prod_{l=0}^{\infty} \left\{1 + y_i^2/(c_i + 2l)^2\right\}^{-1}$, $h(s_0, c_0)$ is a normalising constant, $\theta(z) = \tan^{-1}(z)$, $M(\theta(z)) = (1/2)\log(1 + z^2)$ and $J_\theta(z) = (1 + z^2)^{-1}$. In summary,

$$Z \sim \text{GSSt}(s_0/c_0, c_0),$$
$$Y_i \mid Z \sim \text{GHS}(c_i z, 1/c_i),$$
$$X_i \mid \mathbf{Y}_i^* \sim \text{GSSt}(s_i^*/c_i^*, c_i^*).$$

Some of these conjugate families were presented in Chapter 2; however, there are some likelihoods that were associated to a different conjugate family. The reason is due to a different parameterisation being used.

For example, the negative binomial is usually denoted as $NB(1, \theta)$, where θ is the parameter of success in the Bernoulli trials, but the mean of this distribution is $\mu = (1 - \theta)/\theta$, therefore the mean parameterisation of the negative binomial is obtained if we express θ in terms of the μ, that is, $\theta = 1/(\mu + 1)$. In this case the negative binomial likelihood using the mean parameterisation can be denoted as $NB(1, 1/(\mu + 1))$ as in Table 5.1. According to Section 2.3, the conjugate family for the probability of success θ in a negative binomial distribution is the beta family. If we define the transformation $\mu = (1 - \theta)/\theta$, the induced distribution for μ is an inverse beta distribution, which is the conjugate family for μ in a negative binomial distribution with mean parameterisation.

Another example is the gamma, which is usually denoted as $Ga(1, \theta)$, where θ is the rate parameter, but the mean of this distribution is $\mu = 1/\theta$, therefore the mean parameterisation of the gamma distribution is $Ga(1, 1/\mu)$, as in Table 5.1. According to Section 2.3, the conjugate family for the rate θ in a gamma distribution is another gamma. If we obtain the distribution for the transformation $\mu = 1/\theta$, we obtain the inverse gamma, which is the conjugate family for μ in a gamma distribution with mean parameterisation.

According to Proposition 5.4, we can define dependent sequences with any desired invariant distribution; however, the computations simplify if we use conjugate families. Moreover, if we use any of the six members of the natural exponential family with quadratic variance function (NEF-QVF), the correlation is characterised by the expression given in Proposition 5.5. For conjugate families outside the NEF-QVF family, it is also possible to construct dependence sequences with invariant distribution; however, the calculation of the correlation might not be easy to derive.

For instance, a dependence sequence $\{X_i\}$ with gamma marginal distribution $Ga(a, b)$, can be constructed using conditional distributions based on a Poisson likelihood, given in (i), that is,

$$Z \sim Ga(a, b),$$
$$Y_i \mid Z \sim Po(c_i z),$$
$$X_i \mid \mathbf{Y}_i^* \sim Ga\left(a + \sum_{j \in \partial_i} y_j, b + \sum_{j \in \partial_i} c_j\right).$$

5.3 Particular Exponential Family Cases

Alternatively, it can also be constructed by using a gamma likelihood, as in Section 2.3, in the following way:

$$Z \sim \text{Ga}(a,b),$$
$$Y_i \mid Z \sim \text{Ga}(c_i, z),$$
$$X_i \mid \mathbf{Y}_i^* \sim \text{Ga}\left(a + \sum_{j \in \partial_i} c_j, b + \sum_{j \in \partial_i} y_j\right).$$

The difference is that the former relies on a discrete set of latent variables $\{Y_i\}$ and the correlation can be computed explicitly using Proposition 5.5, whereas in the latter, the construction relies on a set of continuous latent variables $\{Y_i\}$ and the correlation is not easy to compute analytically.

6

Temporal Dependent Sequences

6.1 Moving Average Type of Order q

In this chapter we construct a temporal dependent sequence based on the ideas shown in Section 5.2. The resulting construction is a moving average process of order q. Following Proposition 5.4, the required subsets $\mathbf{Y}_i^*$ to define a $MA(q)$-type process are $\mathbf{Y}_i^* = \{Y_j : j \in \partial_i\}$ with index sets given by

$$\partial_i = \{i - q, \ldots, i - 1, i\} \tag{6.1}$$

for $i = 1, 2, \ldots$, each with $q + 1$ elements, the current index plus q lags. Now, using (6.1), the intersection between two index sets, say ∂_i and ∂_k, is

$$\partial_i \cap \partial_k = \{\max(i, k) - q, \ldots, \min(i, k) - 1, \min(i, k)\}$$

if $|k - i| \le q$. In particular, $\partial_i \cap \partial_k = \emptyset$ if $|k - i| > q$, and $\partial_i \cap \partial_k \ne \emptyset$ if $|k - i| \le q$.

To be specific, let us consider a particular choice of invariant distribution.

Example 6.1 Let the beta model be the invariant distribution. This was originally considered in Jara et al. (2013). To build this model, we use the particular case (ii) from Section 5.3. If we want to use the traditional parameterisation $\text{Be}(a, b)$, we define $a = s_0$ and $b = c_0 - s_0$ and the subsets ∂_i are defined by indexes (6.1). Then

$$s_i^* = s_0 + \sum_{j \in \partial_i} y_j = a + \sum_{j=0}^{q} y_{i-j},$$

$$c_i^* = c_0 + \sum_{j \in \partial_i} c_j = a + b + \sum_{j=0}^{q} c_{i-j}$$

and

$$c_i^* - s_i^* = b + \sum_{j=0}^{q} (c_{i-j} - y_{i-j}).$$

6.1 Moving Average Type of Order q

Therefore, the three-level hierarchical description for the sets of variables $Z, \mathbf{Y} = \{Y_i\}$ and $\mathbf{X} = \{X_i\}$ is

$$Z \sim \text{Be}(a, b),$$
$$Y_i \mid Z \sim \text{Bin}(c_i, z), \qquad (6.2)$$
$$X_i \mid \mathbf{Y} \sim \text{Be}\left(a + \sum_{j=0}^{q} y_{i-j}, b + \sum_{j=0}^{q} (c_{i-j} - y_{i-j})\right)$$

for $i = 1, 2, \ldots$. Proposition 5.4 guarantees that marginally $X_i \sim \text{Be}(a, b)$ for all $i = 1, 2, \ldots$. Moreover, Proposition 5.5 characterises the dependence via the correlation between X_i and X_k as

$$\text{Corr}(X_i, X_k) = \frac{(a + b)\left(\sum_{j=0}^{q-|k-i|} c_{\min(i,k)-j}\right) + \left(\sum_{j=1}^{q} c_{i-j}\right)\left(\sum_{j=1}^{q} c_{k-j}\right)}{\left(a + b + \sum_{j=1}^{q} c_{i-j}\right)\left(a + b + \sum_{j=1}^{q} c_{k-j}\right)}.$$

When $\partial_i \cap \partial_k \neq \emptyset$, if $|k - i| \leq q$, the correlation is higher; however, for $|k - i| > q$, the correlation is lower but does not vanish as in a regular $MA(q)$ model. See Examples 1.9, 5.2 and 5.3.

Let us see some graphical illustrations of the beta process defined in Example 6.1. In Figures 6.1 and 6.2 we show five simulated paths of process $\{X_i\}$ for $i = 1, \ldots, n$ and $n = 20$, on the left side, and autocorrelation function $\rho(l) = \text{Corr}(X_{q+1}, X_{q+1+l})$ as a function of the lag $l = 1, \ldots, 17$, on the right side, for different scenarios.

In Figure 6.1 we defined uniform marginals, obtained by setting $a = b = 1$, and order of dependence $q = 2$, but with different degrees of dependence induced by values of c_i. In row (a) we took $c_i = 0$ to have independent random variables; therefore, the paths of the process X_i freely move, covering the whole interval $[0, 1]$. In row (b) we took $c_i = 1$, which induces a dependence around 0.4; here the paths of the process X_i are slightly less chaotic. Finally, in row (c) we took $c_i = 5$, which induces a correlation of around 0.8. Here the paths of the process X_i stay close to the initial value. In these two latter cases we can see that the correlation is larger for lags with values less than or equal to $q = 2$ and remains constant for lags larger than $q = 2$. This behaviour in the autocorrelation function characterises the $MA(q)$ type of dependence, induced by (6.1) for constant c_i.

On the other hand, instead of having fixed values for c_i, in Figure 6.2, we chose them randomly from Poisson distributions with parameter λ. Additionally, we define marginal beta distributions different from uniforms.

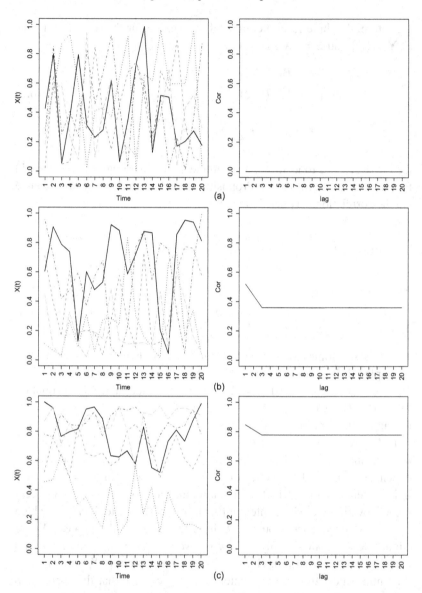

Figure 6.1 Simulations of Example 6.1. Five paths (left side) and autocorrelation function (right side). $a = b = 1$, $q = 2$. $c_i = 0$ (row **(a)**), $c_i = 1$ (row **(b)**) and $c_i = 5$ (row **(c)**).

6.1 Moving Average Type of Order q

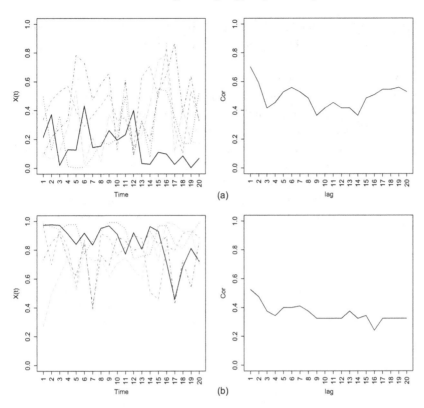

Figure 6.2 Simulations of Example 6.1. Five paths (left side) and autocorrelation function (right side). $q = 2$. $a = 1$, $b = 2$ and $c_i \sim$ Po(2) (row (**a**)), and $a = 5$, $b = 1$ and $c_i \sim$ Po(3) (row (**b**)).

Specifically we took $a = 1$, $b = 2$ and $\lambda = 2$ for row (**a**), and $a = 5$, $b = 1$ and $\lambda = 3$ for row (**b**). Now, in neither of the two cases does the correlation remain constant, so it can increase and decrease for different lags. In row (**a**) the paths are around $1/3 \approx 0.33$, which is the mean of the Be$(1, 2)$ marginals, and in row (**b**) paths are around $5/6 \approx 0.83$, which is the mean of the Be$(5, 1)$ marginals.

In traditional time series analysis (e.g. Box et al. (2015)), the order of dependence q in a moving average process is determined via the sample autocorrelation function defined as

$$ACF(l) = \frac{\sum_{i=1}^{n-l}(X_i - \bar{X})(X_{i+l} - \bar{X})}{\sum_{i=1}^{n}(X_i - \bar{X})^2},$$

where $l = 1,\ldots,n$ is the lag or time shift. This sample autocorrelation function is a point estimate of the autocorrelation function obtained in

Example 1.9. A plot of $ACF(l)$ versus l for $l = 1, \ldots, L$, with L a lot lower than n, has a monotonic decay towards zero, with $ACF(l) \neq 0$ for $l \leq q$ and $ACF(l) = 0$ for $l > q$.

Although for the moving average–type models introduced here, the correlation does not vanish for lags $l > q$, empirical analyses have shown that for n large, a graph of the sample autocorrelation function $ACF(l)$ versus l helps in the detection of the order of dependence q. For the simulated scenarios of Figures 6.1 and 6.2 we extended the sample size to $n = 300$ and computed the sample autocorrelation function $ACF(l)$ for $l = 1, \ldots, 20$. These are shown in Figure 6.3. Recall that in all scenarios the order of dependence was fixed to $q = 2$, but with varying values of dependence parameters c_i. For $c_i = 0$ (panel (a)) the ACF does not detect any order

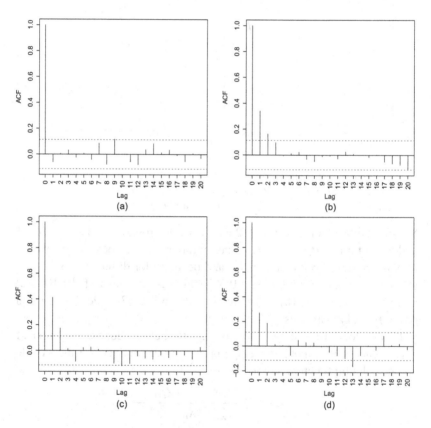

Figure 6.3 ACF of simulated paths of size $n = 300$. In all cases $q = 2$, $a = b = 1$ with $c_i = 0$ (a), $c_i = 1$ (b), $c_i = 5$ (c), and $a = 1$, $b = 2$, $c_i \sim \text{Po}(2)$ (d).

of dependence because, although q was set to the value of 2, when $c_i = 0$ for all i, variables X_i become independent. For $c_i = 1$ and $c_i = 5$ (panels **(b)** and **(c)**) the ACF properly detects the order of dependence $q = 2$ with $ACF(l)$ values outside the confidence intervals for $l \leq 2$ and practically zero for $l > 2$. When c_i is random and taken from a Po(2), although the values of c_i are all different for each i (non-stationary case), the ACF is able to properly detect the order of dependence $q = 2$. This is surprising because the sample ACF estimates the correlation in the stationary case, but in our model with different c_i the process is not stationary.

6.2 Markov versus $MA(q)$ Type

Throughout this book, we have introduced several dependent models with the same invariant distribution, but with different type of dependencies, say exchangeable, Markov or $MA(q)$ type. Apart from the exchangeable model, the latter two models can be used for time series analysis as the probability law of the data. To compare, we use the Be(a, b) as the invariant marginal distribution of the sequences $\{X_i\}$. Specifically, we have

(i) *Markov sequence*: This was defined in Example 4.2 and is constructed as follows:

$$X_1 \sim \text{Be}(a, b), \quad Y_i \mid X_i \sim \text{Bin}(c_i, x_i)$$

and

$$X_{i+1} \mid Y_i \sim \text{Be}(a + y_i, b + c_i - y_i)$$

for $i = 1, 2, \ldots, n$.

(ii) *Moving average type of order q*: This was defined in Example 6.1 and is given by Equations (6.2) and subsets $\mathbf{Y}^*$ defined by set indexes ∂_i given in (6.1).

To see the potential of the different constructions (a) and (b), we carry out a real data analysis.

Unemployment Data Analysis

Consider the monthly unemployment rate in Mexico. Data was obtained from the Mexican National Institute of Geography and Statistics (INEGI) and is given in Tables 3 and 4 of the Appendix. Information is available from January 2006 until July 2023, which gives a total of $n = 211$ months. Since the data is a time series, the subindex i denotes time (month) and the observations X_i the unemployment rate at month $i = 1, \ldots, n$. The data is

90   *Temporal Dependent Sequences*

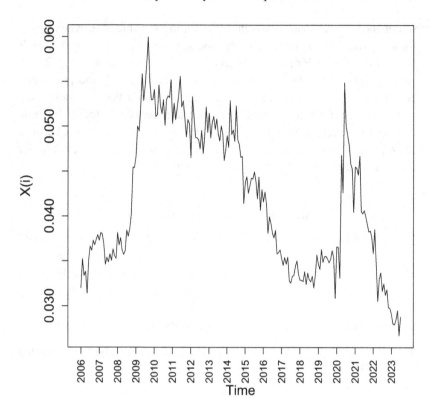

Figure 6.4 Unemployment data set.

shown in Figure 6.4. We see that from 2009 unemployment showed a huge increment, perhaps due to the global mortgage crisis; it remained high until 2015, after which it had a sharp decrease. It remained low from 2017 until the beginning of 2020, where the COVID-19 pandemic appeared in Mexico in March of 2020, causing an acute increment. It was not until 2022 that the rates recovered the values they had before the pandemic with a decreasing tendency.

As an exploratory analysis, we compute the sample autocorrelation function and the sample partial autocorrelation function with the observed data. These are shown in Figure 6.5. The PACF graph suggests an order one or two autoregressive dependence, which implies that the Markov process is going to fit well, and the ACF suggests a long order of dependence for the moving average–type model.

We carry out a Bayesian analysis to fit each of the two models (i) and (ii) to the unemployment data set. In all cases we define vague prior distributions,

6.2 Markov versus $MA(q)$ Type

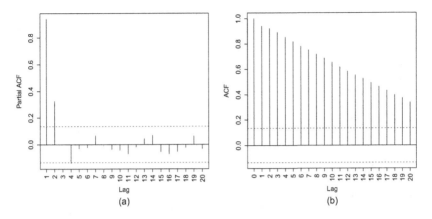

Figure 6.5 PACF **(a)** and ACF **(b)** for unemployment data set.

namely large prior variance, for the model parameters. Specifically, we took independent priors $a \sim \text{Ga}(0.01, 0.01)$, $b \sim \text{Ga}(0.01, 0.01)$, together with a hierarchical prior for the integer dependence parameters $c_i \mid d \stackrel{\text{ind}}{\sim} \text{Po}(d)$ with $d \sim \text{Ga}(1, 1)$ for $i = 1, \ldots, n$ for the Markov and moving average–type models. Finally we took a range of values for $q \in \{1, 2, \ldots, 14\}$ to select the best order of dependence.

Since all the distributions involved in models (i) and (ii) are of standard form, we can run the inference in JAGS (just another Gibbs sampler) (Plummer (2023)). Bugs code for running both models with the given prior specifications can be found in Tables 6.1 and 6.2, respectively.

Before proceeding, we note that the observed data ranges from 0.025 to 0.06, which are very small values. If we run the previous code to the original data, we will see that none of the models behave well. This is due to a numerical precision problem in JAGS. To avoid the numerical problem we re-scale the data to the interval $[0.1, 0.9]$ by applying the linear transformation $X_i^* = 20X_i - 0.3$ and run the models to X_i^* instead.

To compare the fittings we use the two measures $BIAS$ and VAR defined in (3.2). In Figure 6.6 we show the goodness of fit statistics for the $MA(q)$-type models for $q = 1, \ldots, 14$ as solid dotted points and the values of the Markov model as a horizontal dotted line. In panel **(a)** we see that the bias values for the $MA(q)$ models decrease as the value of q increases, stabilising from $q = 5$ and reaching its minimum for $q = 10$. Interestingly, this minimum coincides with the bias obtained by the Markov model. In panel **(b)** we see that the variance's values for the $MA(q)$ models are higher than that of the Markov model for $q \leq 5$ and are smaller for $q \geq 6$. The variance continuously decreases as the value of q increases.

```
model {
#Likelihood
x[1] ~ dbeta(a,b)
for (i in 1:(n-1)) {
y[i] ~ dbin(x[i],c[i]) ax[i] <- a+y[i]
bx[i] <- b+c[i]-y[i]
x[i+1] ~ dbeta(ax[i],bx[i])
}
#Prior
a ~ dgamma(0.01,0.01)
b ~ dgamma(0.01,0.01)
d ~ dgamma(1,1)
for (i in 1:(n-1)) {
c[i] ~ dpois(d)
}
#Prediction
xf[1] ~ dbeta(a,b)
for (i in 1:(n-1)) {xf[i+1] ~ dbeta(ax[i],bx[i])}
}
```

Table 6.1 *Bugs code for implementing Markov model with beta marginals.*

```
model {
#Likelihood
z ~ dbeta(a,b)
for (i in 1:n) {
y[i] ~ dbin(z,c[i])
ax[i] <- a+sum(y[i-0:min(q,i-1)])
bx[i] <- b+sum(c[i-0:min(q,i-1)]-y[i-0:min(q,i-1)])
x[i] ~ dbeta(ax[i],bx[i])
}
#Prior
a ~ dgamma(0.01,0.01)
b ~ dgamma(0.01,0.01)
d ~ dgamma(1,1)
for (i in 1:n) {
c[i] ~ dpois(d)
}
#Prediction
for (i in 1:n) xf[i] ~ dbeta(ax[i],bx[i])
}
```

Table 6.2 *Bugs code for implementing $MA(q)$-type model with beta marginals.*

6.2 Markov versus $MA(q)$ Type

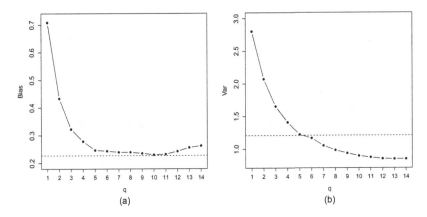

Figure 6.6 Goodness of fit statistics for unemployment data set. *BIAS* (a) and *VAR* (b). $MA(q)$-*type* models with $q = 1, \ldots, 14$ (solid dots) and Markov model (dotted line).

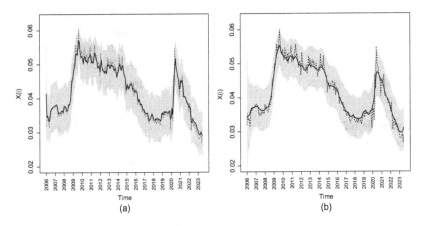

Figure 6.7 Unemployment data set (dotted line). Posterior predictive fits with Markov model (a) and $MA(10)$-type model (b). Point predictions (solid line) and 95% credible intervals (shadows).

Therefore, the Markov model performs well in terms of the bias but not as well in terms of the variances. On the other hand, the best $MA(10)$ model performs as well as the Markov, in terms of the bias, and a lot better in terms of the variances.

To see the performance of the models visually, in Figure 6.7 we present the best fitting of the two different models. The data is shown as a dotted line, the point predictions from the models are shown as solid lines and 95%

credible intervals are depicted as shadows. The Markov model very closely follows the path of the data, but with wider credible intervals whereas the MA-type model, with $q = 10$, very closely follows the tendency of the data, smoothing the path and with narrower credible intervals.

In summary, the two models, Markov and $MA(q)$, are useful for time series modelling with different performances according to what the user is expected to prioritise. Other applications of the $MA(q)$ construction can be found in Nieto-Barajas and Quintana (2016) who proposed a Bayesian nonparametric autoregressive and moving average process based on Pólya trees.

6.3 Seasonal Models

The results given in Propositions 5.4 and 5.5 are very general, so we are able to construct different temporal dependencies apart from the $MA(q)$ type. For instance, Nabeya (2001) proposed seasonal autoregressive models that relate times of different seasons to the behaviour of the current time. For example, if the seasonality of the data is s, we can relate the current time i to previous p seasons, say $\partial_i^{(p)} = \{i-s, i-2s, \ldots, i-ps\}$. This can be achieved by defining subsets $\mathbf{Y}_i^* = \{Y_j : j \in \partial_i^{(p)}\}$. To illustrate, let us consider an example.

Example 6.2 Let us consider again the beta model as the invariant distribution. Then, a seasonal moving average type of model of order p would be defined with sets of variables $Z, \{Y_i\}$, as in the first two equations of (6.2) in Example 6.1, and with $\{X_i\}$ defined as

$$X_i \mid \mathbf{Y} \sim \mathrm{Be}\left(a + \sum_{j=0}^{p} y_{i-js}, b + \sum_{j=0}^{p} (c_{i-js} - y_{i-js})\right). \quad (6.3)$$

Proposition 5.4 shows that $X_i \sim \mathrm{Be}(a, b)$ marginally but with a seasonal dependence of order p. The induced correlation is defined by Proposition 5.5 with $c_0 = a + b$ and $\partial_i^{(p)} \cap \partial_k^{(p)} = \{\max(i, k) - s(p-j), j = 0, \ldots, p-r\}$ if $|i - k| = rs$ with $r = 1, 2, \ldots, p$, and $\partial_i^{(p)} \cap \partial_k^{(p)} = \emptyset$ otherwise.

The seasonal beta process of Example 6.2 is illustrated by taking a seasonality of order $s = 6$ and a dependence of order $p = 2$ with $n = 20$. For uniform marginals, $a = b = 1$, paths of the seasonal process $\{X_i\}$ are shown in the left-side panels in Figure 6.8, together with the correlation function in the right-side panels. Although the marginal distributions are the same in both rows **(a)** and **(b)**, paths in row **(a)** show more variability

6.3 Seasonal Models

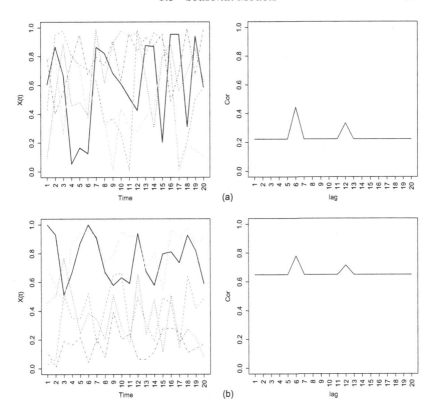

Figure 6.8 Simulations of Example 6.2. Five paths (left side) and autocorrelation function (right side). $a = b = 1$, $q = 2$. $c_i = 1$ (row (**a**)) and $c_i = 5$ (row (**b**)).

due to a smaller dependence than in paths in row (**b**). In both correlation functions there is an increment in the correlation for the two lags 6 and 12 induced by the seasonal construction.

In Figure 6.9, we consider a right skewed beta with parameters $a = 1$ and $b = 2$ and with random parameters $c_i \sim \text{Po}(2)$ in row (**a**), and a left skewed beta with parameters $a = 5$ and $b = 1$ with $c_i \sim \text{Po}(3)$ in row (**b**). Paths of these non-stationary processes are very different with respect to those in Figure 6.8, due to the difference in the invariant marginal distributions, and the correlation functions do not show an apparent increase in lags 6 and 12; however, there is a small peak in both lags.

We also computed the sample autocorrelation function (ACF) for one of the paths of the previous scenarios, but extended the sample size to $n = 300$. This is shown in Figure 6.10. Surprisingly, the ACF manages to detect the

Temporal Dependent Sequences

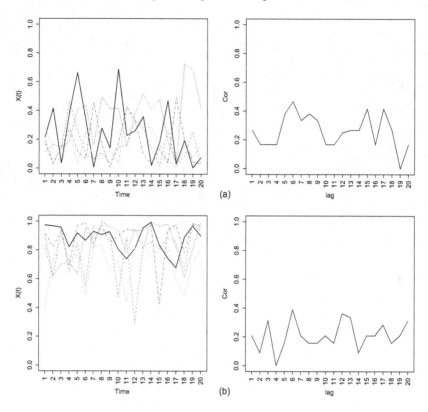

Figure 6.9 Simulations of Example 6.2. Five paths (left-side panels) and autocorrelation function (right-side panels). $q = 2$. $a = 1$, $b = 2$ and $c_i \sim \text{Po}(2)$ (row (**a**)), and $a = 5$, $b = 1$ and $c_i \sim \text{Po}(3)$ (row (**b**)).

appropriate order of dependence with larger values for lags 6 and 12, even for the non-stationary cases where c_i is random. This justifies the use of the sample ACF for exploratory analysis in real data sets.

A more complete model can be obtained by combining the moving average temporal dependence of order q with the seasonal dependence of order p. The required subsets would be $\mathbf{Y}_i^* = \{Y_j : j \in \partial_i^{(q,p)}\}$ with $\partial_i^{(q,p)} = \{i, i-1, \ldots, i-q, i-s, i-2s, \ldots, i-ps\}$, provided $q < s$; otherwise, we would have to be careful not to duplicate the indexes.

To illustrate the performance of seasonal models, let us consider the following real data analysis.

Ozone Data Analysis

Consider monthly maximum tropospheric ozone O_3 concentrations in the metropolitan area of Mexico City. The data was constructed from Mexico

6.3 Seasonal Models

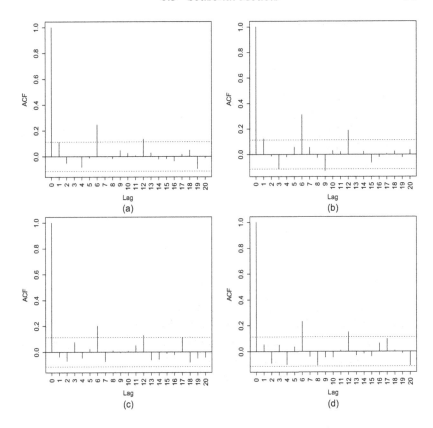

Figure 6.10 ACF of simulated paths of size $n = 300$. In all cases $q = 2$. $a = b = 1$ with $c_i = 0$ (**a**), $c_i = 1$ (**b**), $c_i = 5$ (**c**) and $a = 1$, $b = 2$, $c_i \sim \text{Po}(2)$ (**d**).

City's Atmospheric Monitoring System and is presented in Table 5 of the Appendix. This consists of $n = 98$ observations from January 2007 to February 2015. The data are reported for five regions, but we concentrate on the center region.

The unity of measurement is the metropolitan index of air quality (IMECA), which is a scalar transformation of O_3 concentration levels in parts per million (ppm). This is a scale standardised to easily convey levels of risk to the population. This index takes values between 0 and 300 and can be discretised in five intervals: good $(0, 50]$, regular $(50, 100]$, bad $(100, 150]$, very bad $(150, 200]$ and extremely bad $(200, 300]$ quality of the air.

The data is shown in panel (**a**) in Figure 6.11. We see that most of the time, the IMECA index is between 100 and 150, which means that the quality

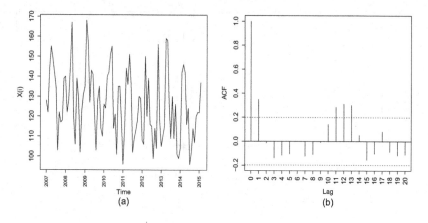

Figure 6.11 Ozone data set. Time series of data **(a)** and ACF **(b)**.

of air is bad most of the time. To see the potential order of dependence in the data, we computed the ACF (panel **(b)** in Figure 6.11). This suggests a one-month order and a seasonal yearly dependence of order 12 months.

Since the the data belongs to a positive scale, we propose a seasonal gamma model and carry out a Bayesian analysis. The model is defined through the following hierarchical representation:

$$Z \sim \text{Ga}(a,b),$$
$$Y_i \mid Z \sim \text{Po}(c_i z),$$
$$X_i \mid \mathbf{Y} \sim \text{Ga}\left(a + \sum_{j=0}^{q} y_{i-j} + \sum_{j=0}^{p} y_{i-js}, b + \sum_{j=0}^{q} c_{i-j} + \sum_{j=0}^{p} c_{i-js}\right).$$

Prior distributions are the same for a and b in the analysis of unemployment data, that is, $a \sim \text{Ga}(0.01, 0.01)$, $b \sim \text{Ga}(0.01, 0.01)$, together with an exchangeable prior for the association parameters $c_i \mid d \sim \text{Ga}(1, d)$ with $d \sim \text{Ga}(1, 1)$ for $i = 1, \ldots, n$. For the MA order of dependence we took a range of values $q \in \{1, 2, 3\}$ with seasonality $s = 12$ and orders $p \in \{0, 1\}$. Bugs code for implementing our model is given in Table 6.3.

To avoid numerical problems, we scaled the data. Specifically, we used $X_i^* = X_i/10$ and ran the model with X_i^*. Model fit was assessed by computing the two goodness of fit measures $BIAS$ and VAR, defined in (3.2) and presented in Table 6.4. The best model is that with moving average order $q = 1$ and seasonal order $p = 1$, as the exploratory analysis suggested. For comparison purposes, we also fitted the Markov gamma model defined as

6.3 Seasonal Models

```
model {
#Likelihood
z ~ dgamma(a,b)
for (i in 1:s) {
muy[i] <- c[i]*z
y[i] ~ dpois(muy[i])
ax[i] <- a+sum(y[i-0:min(q,i-1)])
bx[i] <- b+sum(c[i-0:min(q,i-1)])
x[i] ~ dgamma(ax[i],bx[i])
}
for (i in (s+1):n) {
muy[i] <- c[i]*z
y[i] ~ dpois(muy[i])
ax[i] <- a+sum(y[i-0:min(q,i-1)])+y[i-s]
bx[i] <- b+sum(c[i-0:min(q,i-1)])+c[i-s]
x[i] ~ dgamma(ax[i],bx[i])
}
#Prior
a ~ dgamma(0.01,0.01)
b ~ dgamma(0.01,0.01)
d ~ dgamma(1,1)
for (i in 1:n) {
c[i] ~ dgamma(1,d)
}
#Prediction
for (i in 1:n) {xf[i] ~ dgamma(ax[i],bx[i])}
}
```

Table 6.3 *Bugs code for implementing $MA(q)$-type model plus seasonal component with $p = 1$ and gamma marginals.*

q	p	BIAS	VAR
1	0	132	267
2	0	121	254
3	0	149	262
1	1	**103**	**250**
2	1	114	254
3	1	139	257
Markov		147	263

Table 6.4 *Ozone data set. Goodness of fit measures for different models.*

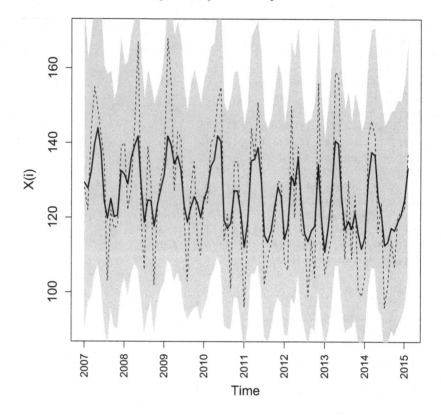

Figure 6.12 Ozone data set (dashed line). Posterior predictive fits moving average seasonal-type model with $q = 1$, $s = 12$ and $p = 1$. Point predictions (solid line) and 95% credible intervals (shadows).

$$X_1 \sim \text{Ga}(a, b),$$
$$Y_i \mid X_i \sim \text{Po}(c_i x_i),$$
$$X_{i+1} \mid Y_i \sim \text{Ga}(a + y_i, b + c_i)$$

for $i = 1, \ldots, n$. Fitting measures show that the moving average model with seasonal component is highly superior. Figure 6.12 shows predictions with the best fitting model. Although the 95% credible intervals are slightly wide, point predictions (solid line) closely follow the path of the observed data (dashed line).

6.4 Periodic Models

Finally, another possible construction for time series analysis are the periodic models. An example of these models for autoregressive type were

6.4 Periodic Models

considered in McLeod (1994). To define them, let us consider monthly data, that is, $s = 12$. Then, every month m could have a different dependence of order q_m. To be clear, let us re-write the index i in terms of year r and month m as $i = i(r,m) = (r-1)s + m$ for $r = 1, 2 \ldots$ and $m = 1, \ldots, s$. In this case, the subsets are $\mathbf{Y}_i^* = \{Y_j : j \in \partial_i^{(q_1,\ldots,q_s)}\}$ with index set $\partial_i^{(q_1,\ldots,q_s)} = \{i(r,m), i(r,m) - 1, \ldots, i(r,m) - q_m\}$. To be specific, let us consider an example.

Example 6.3 For an invariant beta distribution, a periodic moving average type of model of orders $q_1, \ldots, q_s$ would be defined with sets of variables Z, $\{Y_i\}$ as in the first two equations of (6.2) in Example 6.1 and with $\{X_i\}$ defined as

$$X_i \mid \mathbf{Y} \sim \text{Be}\left(a + \sum_{j=0}^{q_m} y_{i(r,s)-j}, b + \sum_{j=0}^{q_m} \left(c_{i(r,s)-j} - y_{i(r,s)-j}\right)\right). \tag{6.4}$$

Proposition 5.4 guarantees that $X_i \sim \text{Be}(a,b)$ marginally but with a periodic dependence of orders $q_1, \ldots, q_m$. The induced correlation is defined by Proposition 5.5 with $c_0 = a+b$ and appropriate definition of the intersection index sets $\partial_i^{(q_1,\ldots,q_m)} \cap \partial_k^{(q_1,\ldots,q_m)}$.

7

Spatial Dependent Sequences

7.1 Latent Areas Model

In the search for temporal dependencies of order larger than one, in Chapter 5, we presented a general result that allows us to keep an invariant distribution for the variables of interest, say X_i, via an appropriate selection of latent variables $\mathbf{Y}_i^*$. In Chapter 6 these subsets of latent variables contained lagged variables, say Y_j for $j \leq i$, because our interest was to define temporal dependencies. However, the result in Proposition 5.4 does not constrain the subsets to be formed of past variables to achieve dependence and keep the marginal distribution invariant. Why not choose a subset that contains future variables, say Y_j for $j > i$? In a time series setting this would not make any sense, but if we move to a spatial setting, we can use the same result to induce spatial dependence.

To be clear, let us consider a hypothetical region with seven areas, like the one depicted in Figure 7.1. In this setting, each area has different sets of neighbours. For instance, area 1 has two neighbours, $\{2, 4\}$, whereas area 5 has five neighbours, $\{2, 3, 4, 6, 7\}$. The graphical representation of spatial dependencies for X_5 in area 5 is shown in Figure 7.2. To avoid overloading the graph with arrows, we only include those for X_5 and the required dependencies between Y_i and X_i, but not the other spatial dependencies for $i \neq 5$.

To define a spatial dependence sequence of variables $\{X_i\}$, let us assume that for each area i there exists a latent variable Y_i for $i = 1, \ldots, n$, plus a common latent variable Z. For each i we define subsets $\mathbf{Y}_i^* = \{Y_j : j \in \partial_i\}$ with index sets given by

$$\partial_i = \{j : j \smile i\} \cup \{i\}, \qquad (7.1)$$

where "$\smile$" denotes spatial neighbour. Then Proposition 5.4 shows us a way to construct a spatial dependence sequence with arbitrary invariant distribution $f_X(x)$. Moreover, if $f_X(x)$ is a member of the exponential family

7.1 Latent Areas Model

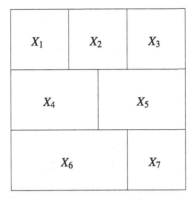

Figure 7.1 Hypothetical region with seven areas.

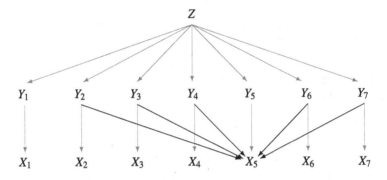

Figure 7.2 Graphical representation of spatial dependencies for X_5 according to the configuration given in Figure 7.1, in the latent areas model.

with quadratic variance, Proposition 5.5 tells us that the correlation between two areas i and k is a function of the the common neighbours $\partial_i \cap \partial_k$.

Going back to the hypothetical region of Figure 7.1, a spatial dependence model would have subsets defined by indexes, $\partial_1 = \{1,2,4\}$ for area 1, $\partial_2 = \{1,2,3,4,5\}$ for area 2, $\partial_3 = \{2,3,5\}$ for area 3, $\partial_4 = \{1,2,4,5,6\}$ for area 4 and $\partial_5 = \{2,3,4,5,6,7\}$ for area 5. Remember that according to (7.1), current index i has to be part of ∂_i. In this case the intersection between index sets, say for regions 4 and 5, would be $\partial_4 \cap \partial_5 = \{2,4,5,6\}$.

Let us consider a particular example.

Example 7.1 Let the beta model be the invariant distribution. This was originally considered in Nieto-Barajas and Hoyos-Argüelles (2024). To build this model, we use the particular case (ii) from Section 5.3. The model is exactly the same as in Example 6.1, but with ∂_i defined by (7.1). That is, the three-level hierarchical description for the sets of variables Z, $\mathbf{Y} = \{Y_i\}$ and $\mathbf{X} = \{X_i\}$ is

$$Z \sim \text{Be}(a, b),$$
$$Y_i \mid Z \sim \text{Bin}(c_i, z), \quad (7.2)$$
$$X_i \mid \mathbf{Y} \sim \text{Be}\left(a + \sum_{j \in \partial_i} y_j, b + \sum_{j \in \partial_i} (c_j - y_j)\right)$$

for $i = 1, 2, \ldots, n$. Proposition 5.4 guarantees that marginally $X_i \sim \text{Be}(a, b)$ for all $i = 1, 2, \ldots, n$. Moreover, Proposition 5.5 characterises the dependence via the correlation between X_i and X_k as

$$\text{Corr}(X_i, X_k) = \frac{(a + b)\left(\sum_{j \in \partial_i \cap \partial_k} c_j\right) + \left(\sum_{j \in \partial_i} c_j\right)\left(\sum_{j \in \partial_k} c_j\right)}{\left(a + b + \sum_{j \in \partial_i} c_j\right)\left(a + b + \sum_{j \in \partial_k} c_j\right)}. \quad (7.3)$$

Now, $\partial_i \cap \partial_k \neq \emptyset$ if areas i and k are neighbours or share a common neighbour, in which case the correlation is higher.

To illustrate how the spatial process defined in Example 7.1 behaves, in Figure 7.3 we show three simulated realisations of a process $\{X_i\}$ based on hierarchical expressions (7.2) and the areas' configuration defined by Figure (7.1). Specifically, we took a uniform marginal distribution, $a = b = 1$, with dependence parameters $c_i \in \{0, 1, 10\}$ to compare. Simulated values x_i were rounded to two decimal places and each of them was associated to a gray intensity where "gray00" is white and "gray99" is black.

According to Figure 7.3, when $c_i = 0$, row (a), each X_i in the path can have totally different intensities because they are independent. For $c_i = 1$, row (b), spatial dependence starts to play a role and intensities show slightly less variation. Finally, for $c_i = 10$, row (c), paths of $\{X_i\}$ are very smooth, with intensities that do not change much. Therefore, in the latent areas model, the role of c_i is to smooth the observed intensities across the different areas.

In the same simulation setting, correlation between any pair can be computed using (7.3). When $c_i = 0$, $\rho_{ik} = 0$ for all $i \neq k = 1, 2, \ldots, 7$ because they are all independent. When $c_i = 1$ for all $i = 1, \ldots, 7$ and $a = b = 1$, the correlations between X_1 and the other areas are

$\rho_{12} = 0.6$, $\rho_{13} = 0.44$, $\rho_{14} = 0.6$, $\rho_{15} = 0.55$, $\rho_{16} = 0.46$, $\rho_{17} = 0.36$.

7.1 Latent Areas Model 105

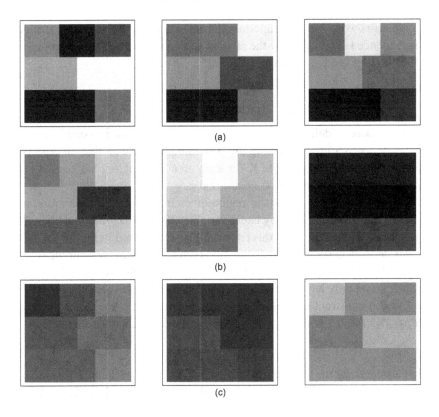

Figure 7.3 Simulations of Example 7.1 with spatial structure (7.1). Three repetitions of the same setting with $a = b = 1$, $c_i = 0$ (row **(a)**), $c_i = 1$ (row **(b)**) and $c_i = 10$ (row **(c)**).

Note that although areas 1 and 7 do not share any neighbour, their correlation is positive, but almost half that of the largest correlation. Finally, when $c_i = 10$ for all i, the correlations between X_1 and the other areas are

$$\rho_{12} = 0.94, \ \rho_{13} = 0.90, \ \rho_{14} = 0.94, \ \rho_{15} = 0.93, \ \rho_{16} = 0.91, \ \rho_{17} = 0.88,$$

which are a lot larger, producing smoother paths.

In this latent areas spatial model, each area i has a spatial parameter c_i that indicates how important the specific area is with respect to its neighbours. Areas with more neighbours have the potential to impact on a larger number of spaces, as compared to areas with less neighbours, which might only have a local impact. However, an area with large c_i but few neighbours will have a constrained local impact, whereas an area with small c_i but with many neighbours won't have an important impact. Therefore in our model spatial dependence is controlled by the number of neighbours and the importance of the area given by parameter c_i.

7.2 Latent Edges Model

Another way of defining a spatially dependent model is by assuming that dependence is different among any two neighbours. For this, instead of considering a latent variable Y_i for each area in a region, we consider a latent edge, say $Y_{i,k}$, that links areas i and k.

Considering the hypothetical region given in Figure 7.1, observation in area 5, X_5, will be defined by latent variables $Y_{2,5}$, $Y_{3,5}$, $Y_{4,5}$, $Y_{5,6}$ and $Y_{5,7}$. A graphical representation of this construction is presented in Figure 7.4. No other dependencies are included in the graph to better convey our ideas.

Let $\{Y_{i,j}\}$ for $i \ne j = 1,\ldots,n$ be a set of latent edges such that $Y_{i,j} = Y_{j,i}$ w.p.1, so we only have at most $n(n-1)/2$ distinct latent variables $\{Y_{i,j}\}$ for $i < j$. To construct a latent edge spatial process $\{X_i\}$, we define subsets $\mathbf{Y}_i^* = \{Y_l : l \in \partial_i\}$ with index sets given by actual neighbours, that is,

$$\partial_i = \{l = (i,j) : j \smile i\}. \quad (7.4)$$

In particular, for the hypothetical region in Figure 7.1, the latent edges index sets would be $\partial_1 = \{(1,2),(1,4)\}$ for area 1, $\partial_2 = \{(2,1),(2,3),(2,4),(2,5)\}$ for area 2, $\partial_3 = \{(3,2),(3,5)\}$ for area 3, $\partial_4 = \{(4,1),(4,2),(4,5),(4,6)\}$ for area 4, $\partial_5 = \{(5,2),(5,3),(5,4),(5,6),(5,7)\}$ for area 5, $\partial_6 = \{(6,4),(6,5),(6,7)\}$ for area 6 and $\partial_7 = \{(7,5),(7,6)\}$ for area 7.

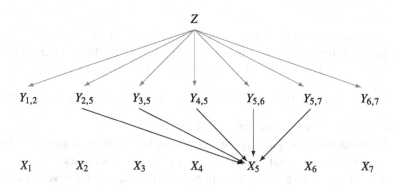

Figure 7.4 Graphical representation of spatial dependencies for X_5 according to the configuration given in Figure 7.1, in the latent edges model.

7.2 Latent Edges Model

If we add a common latent variable Z, Proposition 5.4 shows us a way to construct a spatial dependence model for $\{X_i\}$ with invariant distribution $f_X(x)$. Again, if $f_X(x)$ is a member of the exponential family with quadratic variance, Proposition 5.5 characterises the correlation between any two areas i and k as a function of the common edges $\partial_i \cap \partial_k$, which is the empty set if $i \not\sim k$ or the set $\{(i,k)\}$ if $i \sim k$. So dependence between any two areas is large if the areas are neighbours and small otherwise.

To clearly define our ideas, let us consider a particular example.

Example 7.2 Again, let us consider the beta model as the invariant distribution. This corresponds to the exponential family, in particular case (ii) from Section 5.3. The model is exactly the same as in Example 7.1, but with latent variables $Y_{i,j}$ indexed by a pair of indexes, and ∂_i defined by (7.4). That is, the three-level hierarchical description for the sets of variables Z, $\mathbf{Y} = \{Y_{i,j}\}$ and $\mathbf{X} = \{X_i\}$ is

$$Z \sim \text{Be}(a, b),$$
$$Y_{i,j} \mid Z \sim \text{Bin}(c_{i,j}, z), \qquad (7.5)$$
$$X_i \mid \mathbf{Y} \sim \text{Be}\left(a + \sum_{j \sim i} y_{i,j}, \, b + \sum_{j \sim i} (c_{i,j} - y_{i,j})\right)$$

with $c_{i,j} \equiv c_{j,i}$ for $i, j = 1, 2, \ldots, n$. Proposition 5.4 guarantees that marginally $X_i \sim \text{Be}(a, b)$ for all $i = 1, 2, \ldots, n$. Moreover, Proposition 5.5 characterises the dependence via the correlation between X_i and X_k as

$$\text{Corr}(X_i, X_k) = \frac{(a+b)c_{i,k} I(i \sim k) + \left(\sum_{j \sim i} c_{i,j}\right)\left(\sum_{j \sim k} c_{k,j}\right)}{\left(a + b + \sum_{j \sim i} c_{i,j}\right)\left(a + b + \sum_{j \sim k} c_{k,j}\right)}. \qquad (7.6)$$

Note that $\partial_i \cap \partial_k = \{(i,k)\}$ if areas i and k are neighbours, in which case the correlation is higher, and $\partial_i \cap \partial_k = \emptyset$ if i and k are not neighbours, in which case the correlation is smaller.

We also illustrate the behaviour of the latent edges spatial process defined in Example 7.2. We simulated values for $\{X_i\}$, using the hierarchical representation (7.5) and the neighbouring structure defined by Figure 7.1. In particular, we consider uniform marginal distributions obtained with $a = b = 1$ and different association parameters $c_{i,k} \in \{0, 1, 10\}$ to compare. Again, simulated values x_i were rounded to two decimal places and associated to a greyscale intensity.

Simulated paths are shown in Figure 7.3, where in each row we vary the value of $c_{i,k}$ in $\{0, 1, 10\}$, respectively from top to bottom. More variability is appreciated in paths of the row (a), where the areas' intensities are independent. As we increase the value of $c_{i,k}$, rows (b) and (c), the paths become smoother.

As in the preceding simulation scenario, theoretical correlations between any pair of random variables, say X_i and X_k, can be computed using (7.6). In particular, for $c_{i,k} = 0$, $\rho_{i,k} = 0$ for all i and k since they become independent. For $a = b = 1$ and $c_{i,k} = 1$ for all i and k, the correlations between X_1 and the rest of the areas are

$$\rho_{12} = 0.42, \; \rho_{13} = 0.25, \; \rho_{14} = 0.42, \; \rho_{15} = 0.36, \; \rho_{16} = 0.3, \; \rho_{17} = 0.25.$$

Again, although areas 1 and 7 are not neighbours, their correlations are positive, but have the same values as the correlations between areas 1 and 3. Finally, for $c_{i,k} = 10$ for all i and k, the correlations between X_1 and the other areas are

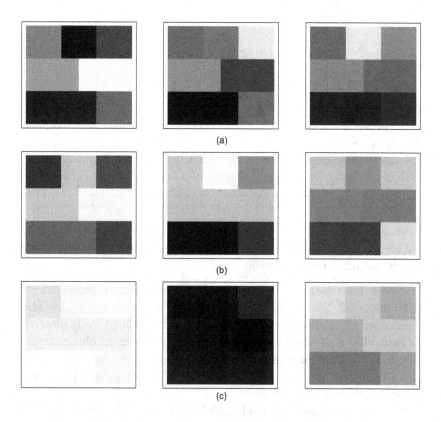

Figure 7.5 Simulations of Example 7.2 with spatial structure (7.4). Three repetitions of the same setting with $a = b = 1$, $c_{i,j} = 0$ (row **(a)**), $c_{i,j} = 1$ (row **(b)**) and $c_{i,j} = 10$ (row **(c)**).

$\rho_{12} = 0.89$, $\rho_{13} = 0.83$, $\rho_{14} = 0.89$, $\rho_{15} = 0.87$, $\rho_{16} = 0.85$, $\rho_{17} = 0.83$.

We can see that these numbers are slightly smaller than those obtained by the latent areas model for the same values of $c_i = c_{i,k}$.

One difference between the two previous spatial models is the number of parameters. In the latent areas model, the number of dependence parameters $\{c_i\}$ is n, one for each area in the region. On the other hand, in the latent edges model, the number of parameters $\{c_{i,k}\}$ is something between $n-1$ and $n(n-1)/2$, depending on the spatial configuration of the areas. For instance, considering the region given in Figure 7.1, the number of dependence parameters in the latent areas model is 7, whereas in the latent edges model it is 13.

7.3 Model Comparison

In the previous two sections we have presented two types of spatial dependent models, one based on latent areas and the other based on latent edges. In order to further study their differences, we carry out a data analysis.

Unemployment Data Analysis

Consider the third quarter of 2023 unemployment rate in Mexico for the 32 states of the country. Data was obtained from INEGI and is given in Table 6 of the Appendix. In this case data consists of areal (states') observations, the subindex i denotes a state and the observations X_i the unemployment rate at state $i = 1, \ldots, n$ with $n = 32$. The data is shown in Figure 7.6 (a). States with higher unemployment rates are two in the north, Chihuahua (ID 5) and Coahuila (ID 28); three more in the metropolitan area (center of the country), Distrito Federal (ID 9), State of Mexico (ID 15) and Tlaxcala (ID 29); plus one more in the south-east, Tabasco (ID 27). Overall, by looking at the whole country, the northern part of Mexico shows the higher unemployment rates. Surprisingly, the states with the lowest unemployment rate are the poorest states, Guerrero (ID 12), Oaxaca (ID 20) and Michoacan (ID 16), all situated in the south of the country.

We carried out a Bayesian analysis to fit the latent areas and latent edges models of Sections 7.1 and 7.2, respectively. In particular, we assume a beta marginal distribution for the rates and use constructions outlined in Examples 7.1 and 7.2. We took vague prior distributions for the model parameters. Specifically, $a \sim \text{Ga}(0.01, 0.01)$, $b \sim \text{Ga}(0.01, 0.01)$, together with hierarchical priors for dependence parameters $c_i \mid d \sim \text{Po}(d)$ and $d \sim \text{Ga}(5, 1)$ for the latent areas model and $c_{ij} \mid d \sim \text{Po}(d)$ and $d \sim \text{Ga}(5, 1)$ for the latent edges model.

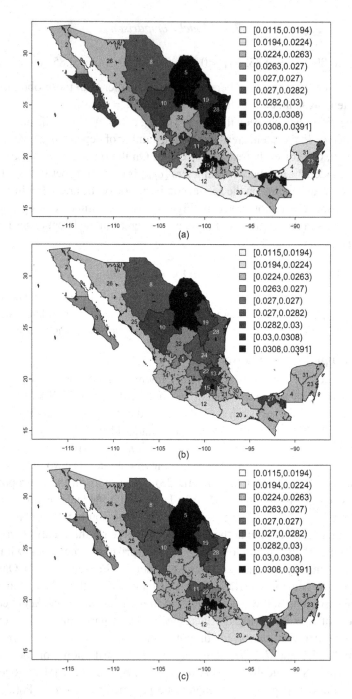

Figure 7.6 Unemployment rates for the 32 states of Mexico during the 3rd quarter of 2023. Observed data (**a**), latent areas fit (**b**) and latent edges fit (**c**). Legends have been kept the same for the three maps.

7.3 Model Comparison

The neighbourhood structure of the states of Mexico is available in Table 7 of the Appendix, which is needed to identify when $i \smile j$. With this information we can define the subsets (7.1) and (7.4) required in the definition of latent areas and latent edges models, respectively.

Since all distributions involved in both models are of standard form, we run the inference in JAGS (Plummer (2023)). Bugs code for running both models with the given prior specifications can be found in Tables 7.1 and 7.2, respectively.

As in the analysis of unemployment data of Section 6.2, we transform the data from the observed values $[0.01, 0.04]$ to $[0.1, 0.9]$ via the linear transformation $X_i^* = 80 X_i/3 - 1/6$ and run the model to X_i^*. Again, we compare the fittings using the two measures $BIAS$ and VAR defined in (3.2). According to Table 7.3, the bias is smaller for the latent edges model, which means that the latent edges model smooth the observed data less than the latent areas model. On the other hand, the latent areas model has slightly smaller variance, perhaps due to a larger number of parameters shared in the conditional beta distributions.

```
model {
#Likelihood
z ~ dbeta(a,b)
for (i in 1:n) {
y[i] ~ dbin(z,c[i])
ax[i] <- a+sum(y[]*w[i,])
bx[i] <- b+sum((c[]-y[])*w[i,])
x[i] ~ dbeta(ax[i],bx[i])
}
#Prior
a ~ dgamma(0.01,0.01)
b ~ dgamma(0.01,0.01)
d ~ dgamma(5,1)
for (i in 1:n) {
c[i] ~ dpois(d)
}
#Prediction
for (i in 1:n) {xf[i] ~ dbeta(ax[i],bx[i])}
}
```

Table 7.1 *Bugs code for implementing latent areas model with beta marginals.*

```
model {
#Likelihood
z ~ dbeta(a,b)
for (i in 1:n) {
for (j in (i+1):n) {
y[i,j] ~ dbin(z,c[i,j])
}
}
ax[1] <- a+sum(y[1,2:n]*w[1,2:n])
bx[1] <- b+sum((c[1,2:n]-y[1,2:n])*w[1,2:n])
x[1] ~ dbeta(ax[1],bx[1])
for (i in 2:(n-1)) {
ax[i] <- a+sum(y[1:(i-1),i]*w[1:(i-1),i])
         +sum(y[i,(i+1):n]*w[i,(i+1):n])
bx[i] <- b+sum((c[1:(i-1),i]-y[1:(i-1),i])*w[1:(i-1),i])
         +sum((c[i,(i+1):n]-y[i,(i+1):n])*w[i,(i+1):n])
x[i] ~ dbeta(ax[i],bx[i])
}
ax[n] <- a+sum(y[1:(n-1),n]*w[1:(n-1),n])
bx[n] <- b+sum((c[1:(n-1),n]-y[1:(n-1),n])*w[1:(n-1),n])
x[n] ~ dbeta(ax[n],bx[n])
#Prior
a ~ dgamma(0.01,0.01)
b ~ dgamma(0.01,0.01)
d ~ dgamma(5,1)
for (i in 1:(n-1)) {
for (j in (i+1):n) {
c[i,j] ~ dpois(d)
}
}
#Prediction
for (i in 1:n) {xf[i] ~ dbeta(ax[i],bx[i])}
}
```

Table 7.2 *Bugs code for implementing latent edges model with beta marginals.*

The actual benefit of spatial models is that they borrow strength across neighbours and produce smoother predictions for each of the areas. In Figure 7.6 we also show the predictions obtained with each of the two spatial models. The latent areas model (row (**b**)) produces a map that is smoother than the one produced by the latent edges model (row (**c**)) in the sense that colour transitions are smoother across states. On the other hand, the latent edges model still identifies those states in the center of the country with a distinctive (higher) unemployment rate compared to their

Model	BIAS	VAR
Latent areas	0.0005672	0.001132
Latent edges	0.0003722	0.001139

Table 7.3 *Spatial unemployment data set. Goodness of fit measures for different models.*

neighbours. By visually comparing the three maps, we see that maps (**b**) and (**c**) are smooth versions of the observed data (map (**a**)), with the latent edges model producing a low degree of smoothness and the latent areas model producing a high degree of smoothness. They are both useful and which one is better is a matter of taste for the decision maker.

This example illustrates the use of the spatial constructions to model data. However, they can also be used as spatial random effects. See, for example, Nieto-Barajas (2008) who analysed disease mapping data, and Nieto-Barajas and Bandyopadhyay (2013) in the context of zero-inflated models.

7.4 Spatio-Temporal Models

In Chapter 6 we showed how to use the results stated in Chapter 5 to construct temporal dependence models. These temporal constructions can be combined with either of the two spatial models presented in Sections 7.1 and 7.2. to define spatio-temporal models. Moreover, the result in Proposition 5.4 guarantees that the marginal distribution is invariant, even in the spatio-temporal setting.

Imagine that for each region $i = 1,\ldots,n$ we have measured random variable $X_{t,i}$ at different time points, say $t = 1,\ldots,T$. Considering the latent areas setting of Section 7.1, suppose that for each region i and time t we have a latent variable $Y_{t,i}$ and a common variable Z.

A spatio-temporal dependence model for $X_{t,i}$ would link latent variables $Y_{t,i}$ from neighbouring regions j such that $j \smile i$ and from previous times s such that $s < t$, together with the current index (t,i). Specifically, the subsets $\mathbf{Y}^*_{t,i} = \{Y_{s,j} : (s,j) \in \partial_{t,i}\}$ would be defined by index sets

$$\partial_{t,i} = \{(s,j): s \in \{t-q,\ldots,t-1,t\}\} \cup \{(t,j): j \smile i\}. \qquad (7.7)$$

Let us consider a region like the one depicted in Figure 7.1. A graphical representation of a spatio-temporal dependence model for area $i = 5$ at time $t = 3$ with $q = 2$, using (7.7), is shown in Figure 7.7. Dependencies for other regions at different times, as well as dependencies between latents $Y_{2,i}$ and $Y_{3,i}$ on Z, are not included.

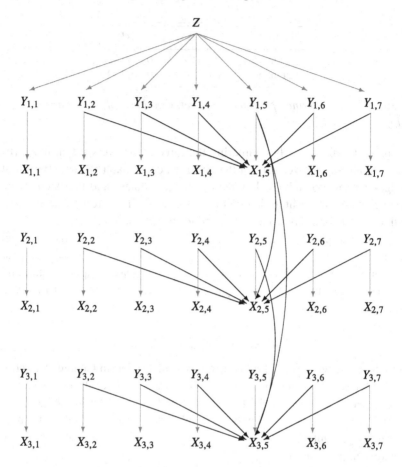

Figure 7.7 Graphical representation of spatio-temporal dependencies for $X_{1,5}$, $X_{2,5}$ and $X_{3,5}$ according to the spatial configuration given in Figure 7.1 and for $q = 2$, based on the latent areas model. All random variables $Y_{t,i}$ are conditionally independent given Z, but not all arrows are shown.

Within the same setting, an alternative spatio-temporal construction can be defined by considering that the spatial neighbours are neighbours for all lagged times, that is, the index sets would be

$$\partial_{t,i} = \{(s,j): s \in \{t-q,\ldots,t-1,t\};\ j \smile i, j = i\}. \tag{7.8}$$

These and other types of spatio-temporal dependencies were studied: for the normal model in Nieto-Barajas (2020) in the context of generalised linear models with space-time varying coefficients; for the Pareto model in Nieto-Barajas and Huerta (2017) in the analysis of heavy tail pollution

7.4 Spatio-Temporal Models

data; and for the beta model in Nieto-Barajas (2022a) in the context of evolutionary mortality rates.

In the following example we present a spatio-temporal model that considers dynamic latent areas and a specific marginal distribution.

Example 7.3 Let us consider the beta model as the invariant distribution. The model uses the exponential family particular case (ii) from Section 5.3, and the latent variables $Y_{t,i}$ are indexed by a pair of indexes (t,i) which refer to time and space, respectively. The index sets would be either (7.7) or (7.8). The three-level hierarchical description for the sets of variables Z, $\mathbf{Y} = \{Y_{t,i}\}$ and $\mathbf{X} = \{X_{t,i}\}$ is

$$Z \sim \text{Be}(a,b),$$
$$Y_{t,i} \mid Z \sim \text{Bin}\left(c_{t,i}, z\right), \tag{7.9}$$
$$X_{t,i} \mid \mathbf{Y} \sim \text{Be}\left(a + \sum_{(s,j) \in \partial_{t,i}} y_{s,j}, \, b + \sum_{(s,j) \in \partial_{t,i}} (c_{s,j} - y_{s,j})\right)$$

for $t = 1,\ldots,T$ and $i = 1,2,\ldots,n$. Proposition 5.4 guarantees that marginally $X_{t,i} \sim \text{Be}(a,b)$ for all (t,i). Moreover, Proposition 5.5 characterises the dependence via the correlation between $X_{t,i}$ and $X_{r,k}$ as

$$\text{Corr}(X_{t,i}, X_{r,k}) = \frac{(a+b)\sum_{(s,j) \in \partial_{t,i} \cap \partial_{r,k}} c_{s,j} + \left(\sum_{(s,j) \in \partial_{t,i}} c_{s,j}\right)\left(\sum_{(s,j) \in \partial_{r,k}} c_{s,j}\right)}{\left(a + b + \sum_{(s,j) \in \partial_{t,i}} c_{s,j}\right)\left(a + b + \sum_{(s,j) \in \partial_{r,k}} c_{s,j}\right)}.$$

The correlation is higher between time-space locations (t,i) and (r,k) that share more latent areas and times.

Another way of defining a space-time dependent model for $\{X_{t,i}\}$ is by extending the latent edges model of Section 7.2 to consider time. Imagine that for each time $t = 1,\ldots,T$, there is a latent edge $Y_{t,i,j}$ that links areas i and j, with $Y_{t,j,i} = Y_{t,i,j}$ w.p.1, for $i,j = 1,\ldots,n$. Additionally, we need a common variable Z.

A spatio-temporal dependence model for $X_{t,i}$ would link latent spatio-temporal edges $Y_{t,i,j}$ from neighbouring regions j such that $j \smile i$ and from previous times s such that $s < t$. Specifically, the subsets $\mathbf{Y}^*_{t,i} = \{Y_{s,i,j} : (s,i,j) \in \partial_{t,i}\}$ would be defined by index sets

$$\partial_{t,i} = \{(s,i,j) : s \in \{t-q,\ldots,t-1,t\}; \, j \smile i\}. \tag{7.10}$$

Let us consider a region like the one depicted in Figure 7.1. A graphical representation of a spatio-temporal dependence model for area $i = 5$ at time $t = 2$ with $q = 1$, using (7.10), is shown in Figure 7.8. Dependencies for other regions at different times and between $Y_{2,i,j}$ on Z, are not included.

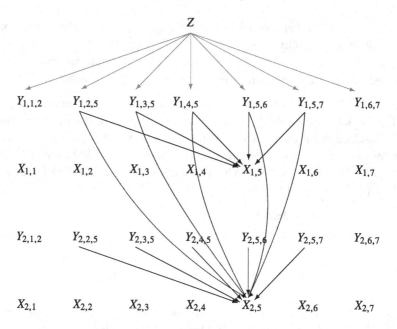

Figure 7.8 Graphical representation of spatio-temporal dependencies for $X_{1,5}$ and $X_{2,5}$ according to the spatial configuration given in Figure 7.1 and for $q = 1$, based on the latent edges model. All random variables $Y_{t,i,k}$ are conditionally independent, given Z, but not all arrows are shown.

A particular example for a spatio-temporal construction based on latent edges is given as follows.

Example 7.4 To be consistent with the previous example, let us consider the beta model as the invariant distribution. The model uses the exponential family particular case (ii) from Section 5.3 and the latent variables $Y_{t,i,j}$ are indexed by three indexes (t,i,j) which refer to time t and edge (i,j) between locations i and j. The index sets would be like in (7.10). The three-level hierarchical description for the sets of variables Z, $\mathbf{Y} = \{Y_{t,i,j}\}$ and $\mathbf{X} = \{X_{t,i}\}$ is

$$Z \sim \text{Be}(a,b),$$
$$Y_{t,i,j} \mid Z \sim \text{Bin}(c_{t,i,j}, z), \qquad (7.11)$$
$$X_{t,i} \mid \mathbf{Y} \sim \text{Be}\left(a + \sum_{s=0}^{q}\sum_{j \sim i} y_{t-s,i,j},\ b + \sum_{s=0}^{q}\sum_{j \sim i}(c_{t-s,i,j} - y_{t-s,i,j})\right)$$

7.4 Spatio-Temporal Models

for $t = 1,\ldots,T$ and $i = 1,2,\ldots,n$. Proposition 5.4 guarantees that marginally $X_{t,i} \sim \text{Be}(a,b)$ for all (t,i). Moreover, Proposition 5.5 characterises the dependence via the correlation between $X_{t,i}$ and $X_{r,k}$ as

$$\frac{(a+b)\sum_{s=0}^{q-|t-r|} c_{\min(t,r)-s,i,k} I(i \smile k) +}{\left(a+b+\sum_{s=0}^{q}\sum_{j\smile i} c_{t-s,i,j}\right)\left(a+b+\sum_{s=0}^{q}\sum_{j\smile k} c_{r-s,k,j}\right)}.$$

The correlation is higher between times t and r such that $|t - r| \leq q$ and between locations i and k such that $i \smile k$.

8

Multivariate Dependent Sequences

8.1 Multivariate Models

Throughout this book we have proposed several dependent models based on hierarchical constructions to achieve a desired marginal distribution. Dependences include exchangeable, Markov, moving average, spatial, spatio-temporal, periodic and seasonal.

All constructions are based on an appropriate use of four building blocks, say $f_X(x)$, $f_{X|Y}(x \mid y)$, $f_{Y|X}(y \mid x)$ and $f_Y(y)$. We usually rely on conjugate families, but non-conjugate families can also be used as long as the respective densities are available in close form.

All examples presented in the previous chapters deal with univariate sequences; however, we can also create dependence sequences of random vectors. Typical choices are the Dirichlet and multivariate normal, which are conjugate models for the multinomial and multivariate normal, respectively. Here we present these two cases in detail.

Let us consider the Dirichlet case first.

Example 8.1 Our building blocks for constructing dependence sequences come from the Dirichlet and multinomial conjugate family. Let $\mathbf{X}_i = (X_{i,1}, \ldots, X_{i,k})$ and $\mathbf{Y} = (Y_1, \ldots, Y_k)$ be random vectors of dimension $k > 0$. To construct an exchangeable sequence $\{\mathbf{X}_i\}$ whose marginal distribution is Dirichlet, we take

$$\mathbf{Y} \sim \text{DMult}(\mathbf{a}, c) \quad \text{or} \quad \mathbf{Z} \sim \text{Dir}(\mathbf{a}) \text{ with } \mathbf{Y} \mid \mathbf{Z} \sim \text{Mult}(c, \mathbf{z})$$

and

$$\mathbf{X}_i \mid \mathbf{Y} \sim \text{Dir}(\mathbf{a} + \mathbf{y})$$

for $i = 1, \ldots, n$. According to Proposition 3.5, $\mathbf{X}_i \sim \text{Dir}(\mathbf{a})$ marginally. Dependence measures between any vector components $X_{i,r}$ and $X_{j,s}$ of vectors $\mathbf{X}_i$ and $\mathbf{X}_j$ can be obtained explicitly. For $r = s$, that is, the same vector component, marginal distributions reduce to the beta binomial conjugate family, therefore

118

8.1 Multivariate Models

$$\text{Cov}(X_{i,r}, X_{j,r}) = \text{Cov}\{E(X_{i,r} \mid Y_r), E(X_{j,r} \mid Y_r)\}$$
$$= \frac{\text{Var}(Y_r)}{(a_0 + c)^2} = \frac{c a_r (a_0 - a_r)}{(a_0 + c) a_0^2 (a_0 + 1)},$$

where $a_0 = \sum_{r=1}^{k} a_r$. Now, since $X_{i,r} \sim \text{Be}(a_r, a_0 - a_r)$, the correlation becomes

$$\text{Corr}(X_{i,r}, X_{j,r}) = \frac{c}{a_0 + c}. \tag{8.1}$$

For $r \neq s$, we get

$$\text{Cov}(X_{i,r}, X_{j,s}) = \text{Cov}\{E(X_{i,r} \mid \mathbf{Y}), E(X_{j,s} \mid \mathbf{Y})\}$$
$$= \frac{\text{Cov}(Y_r, Y_s)}{(a_0 + c)^2} = -\frac{c(a_0 + c) a_r a_s}{(a_0 + c)^2 a_0^2 (a_0 + 1)}.$$

Again, using the marginal beta distributions, the correlation becomes

$$\text{Corr}(X_{i,r}, X_{j,s}) = -\left(\frac{c}{a_0 + c}\right) \sqrt{\frac{a_r a_s}{(a_0 - a_r)(a_0 - a_s)}}. \tag{8.2}$$

If we want to construct a Markov process $\{\mathbf{X}_i\}$ with Dirichlet as the invariant distribution, we take

$$\mathbf{X}_1 \sim \text{Dir}(\mathbf{a}), \quad Y_i \mid \mathbf{X}_i \sim \text{Mult}(c_i, \mathbf{x}_i)$$

and

$$\mathbf{X}_{i+1} \mid \mathbf{Y}_i \sim \text{Dir}(\mathbf{a} + \mathbf{y}_i)$$

for $i = 1, \ldots, n - 1$. According to Proposition 4.1, $\mathbf{X}_i \sim \text{Dir}(\mathbf{a})$ marginally. The correlations between $X_{i,r}$ and $X_{i+1,s}$ are the same as in (8.1) and (8.2) but with c replaced by c_i.

Moreover, if we want to construct a temporal or spatial sequence $\{\mathbf{X}_i\}$ whose invariant distribution is Dirichlet, we take

$$\mathbf{Z} \sim \text{Dir}(\mathbf{a}), \quad Y_i \mid \mathbf{Z} \sim \text{Mult}(c_i, \mathbf{z})$$

and

$$\mathbf{X}_i \mid \mathbf{Y}_i^* \sim \text{Dir}\left(\mathbf{a} + \sum_{j \in \partial_i} \mathbf{y}_j\right)$$

for an appropriate choice of subsets $\mathbf{Y}_i^*$ and index sets ∂_i for $i = 1, \ldots, n$. According to Proposition 5.4, $\mathbf{X}_i \sim \text{Dir}(\mathbf{a})$. Correlation between $X_{i,r}$ and $X_{k,s}$, for $r = s$, is given in Proposition 5.5 and has the expression

$$\text{Corr}(X_{i,r}, X_{k,r}) = \frac{a_0 \sum_{j \in \partial_i \cap \partial_k} c_j + \left(\sum_{j \in \partial_i} c_j\right)\left(\sum_{j \in \partial_k} c_j\right)}{\left(a_0 + \sum_{j \in \partial_i} c_j\right)\left(a_0 + \sum_{j \in \partial_k} c_j\right)}, \tag{8.3}$$

and for $r \neq s$, it becomes

$$\text{Corr}(X_{i,r}, X_{k,s}) = -\sqrt{\frac{a_r a_s}{(a_0 - a_r)(a_0 - a_s)}} \text{Corr}(X_{i,r}, X_{k,r}).$$

In all cases, the strength of the dependence is controlled by the single parameter c in the exchangeable case and the parameters $\{c_i\}$ in the Markov and general dependence cases.

Applications of dependent Dirichlet sequences have been considered in Bekele et al. (2012) to study the dynamics in contingency tables for patients with progressive cancer stages. Likewise, they have been studied by Pan et al. (2024) for the study of temporal dependent mixtures of copulas.

We now describe the normal case.

Example 8.2 Our building blocks for constructing dependence sequences come from the multivariate normal and multivariate normal conjugate family. To construct an exchangeable sequence $\{\mathbf{X}_i\}$ whose marginal distribution is multivariate normal, we take

$$\mathbf{Y} \sim N_p(\boldsymbol{\mu}, (C^{-1} + A^{-1})^{-1}) \quad \text{or} \quad \mathbf{Z} \sim N_p(\boldsymbol{\mu}, A) \quad \text{with} \quad \mathbf{Y} \mid \mathbf{Z} \sim N_p(\mathbf{z}, C)$$

and

$$\mathbf{X}_i \mid \mathbf{Y} \sim N_p\left((A + C)^{-1}(A\boldsymbol{\mu} + C\mathbf{y}), A + C\right)$$

for $i = 1, \ldots, n$. According to Proposition 3.5, $\mathbf{X}_i \sim N_p(\boldsymbol{\mu}, A)$ marginally. Dependence measures between any two vectors $\mathbf{X}_i$ and $\mathbf{X}_j$ can be computed explicitly. In particular, the covariance matrix is given by

$$\text{Cov}(\mathbf{X}_i, \mathbf{X}_j) = \text{Cov}\{E(\mathbf{X}_i \mid \mathbf{Y}), E(\mathbf{X}_j \mid \mathbf{Y})\} = \text{Var}\{(A + C)^{-1}(A\boldsymbol{\mu} + C\mathbf{Y})\}$$
$$= (A + C)^{-1}C(C^{-1} + A^{-1})C(A + C)^{-1},$$

which, after considering that $I = AA^{-1} = CC^{-1}$, simplifies to

$$\text{Cov}(\mathbf{X}_i, \mathbf{X}_j) = (A + C)^{-1}CA^{-1}. \tag{8.4}$$

It is not straightforward to see that this matrix is symmetric, but by multiplying the covariance matrix and its transpose, on the left and right sides, by $(A + C)$, we get the same result, which implies that the covariance and its transpose must be equal.

Now, if we want to construct a Markov process $\{\mathbf{X}_i\}$ with multivariate normal as the invariant distribution, we take

$$\mathbf{X}_1 \sim N_p(\boldsymbol{\mu}, A), \quad \mathbf{Y}_i \mid \mathbf{X}_i \sim N_p(\mathbf{x}_i, C_i)$$

8.1 Multivariate Models

and

$$X_{i+1} \mid Y_i \sim N_p\left((A + C_i)^{-1}(A\mu + C_i y_i), A + C_i\right)$$

for $i = 1, \ldots, n-1$. According to Proposition 4.1, $X_i \sim N_p(\mu, A)$ marginally. The covariance matrix between X_{i+1} and X_i is the same as (8.4) but with C replaced by C_i. Moreover, the covariance between X_{i+2} and X_i, using the reversibility of the construction and the iterative law of the covariance, we get

$$Cov(X_{i+2}, X_i) = Cov\{(A + C_{i+1})^{-1}(A\mu + C_{i+1} Y_{i+1}), (A + C_i)^{-1}(A\mu + C_i Y_i)\}$$
$$= (A + C_{i+1})^{-1} C_{i+1} Cov(Y_{i+1}, Y_i) C_i (A + C_i)^{-1}.$$

Now, since $Cov(Y_{i+1}, Y_i) = Var(X_{i+1}) = A^{-1}$, the two-lag covariance becomes

$$Cov(X_{i+2}, X_i) = (A + C_{i+1})^{-1} C_{i+1} A^{-1} C_i (A + C_i)^{-1}.$$

Finally, if we want to construct a temporal or spatial sequence $\{X_i\}$ whose invariant distribution is multivariate normal, we take

$$Z \sim N_p(\mu, A), \quad Y_i \mid Z \sim N_p(z, C_i)$$

and

$$X_i \mid Y_i^* \sim N_p\left(\left(A + \sum_{j \in \partial_i} C_j\right)^{-1} \left(A\mu + \sum_{j \in \partial_i} C_j y_j\right), A + \sum_{j \in \partial_i} C_j\right)$$

for an appropriate choice of subsets Y_i^* and index sets ∂_i for $i = 1, \ldots, n$. According to Proposition 5.4, $X_i \sim N_p(\mu, A)$. The covariance matrix between X_i and X_k, after using the iterative covariance formula twice, is given by

$$Cov(X_i, X_k) = \left(A + \sum_{j \in \partial_i} C_j\right)^{-1}$$
$$\times \left\{\sum_{j \in \partial_i \cap \partial_k} C_j + \left(\sum_{j \in \partial_i} C_j\right) A^{-1} \left(\sum_{j \in \partial_k} C_j\right)\right\} \left(A + \sum_{j \in \partial_k} C_j\right)^{-1}.$$

In all cases, the strength of the dependence is controlled by the single parameter matrix C in the exchangeable case and the parameters' matrices $\{C_i\}$ in the Markov and general dependence cases.

122   *Multivariate Dependent Sequences*

8.2 Dirichlet Process Models

Apart from the two multivariate cases we described earlier, we can go beyond and construct dependence sequences of random measures. To illustrate, we consider the Dirichlet process introduced in Section 3.3.

Let $G \sim \mathcal{DP}(a, F)$, where $a > 0$ is the precision parameter and F is the centring measure. One way of characterising a Dirichlet process is through its finite-dimensional distributions. That is, for any integer $K > 0$ and any partition $(B_1, \ldots, B_K)$ of $\mathbb{R}$, we have that $(G(B_1), \ldots, G(B_K)) \sim \text{Dir}(aF(B_1), \ldots, aF(B_K))$. On the other hand, we define a multinomial process N with integer parameter c and probability measure F, in notation $N \sim \mathcal{MP}(c, F)$. Again, its finite-dimensional distributions for any $K > 0$ and partition $(B_1, \ldots, B_K)$ of $\mathbb{R}$ are $(N(B_1), \ldots, N(B_K)) \sim \text{Mult}(c, F(B_1), \ldots, F(B_K))$. Alternatively, if we sample c random variables $Y_1, \ldots, Y_c$ independently and identically distributed from F, then the multinomial process is re-written as $N(\cdot) = \sum_{i=1}^{c} \delta_{Y_i}(\cdot) = c\widehat{G}$, where δ_Y is the Dirac delta and $\widehat{G}$ is the empirical distribution function for the Y_i.

Moreover, if conditionally on G we have that $N \mid G \sim \mathcal{MP}(c, G)$ together with $G \sim \mathcal{DP}(a, F)$, then marginally $N \sim \mathcal{DMP}(a, c, F)$, that is, a Dirichlet-multinomial process with scalar parameters a, c and probability measure parameter F. Additionally, it can be proved that $G \mid N \sim \mathcal{DP}(a + c, (aF + N)/(a + c))$. Note that the centring measure of this conditional distribution can be re-written as $(aF + c\widehat{G})/(a + c)$, which defines a mixture between F and $\widehat{G}$. See Walker and Muliere (2003) for details.

We are now in a position to define dependence sequences of Dirichlet processes.

Example 8.3   Our building blocks are the Dirichlet, multinomial and Dirichlet-multinomial processes. To construct an exchangeable sequence $\{F_i\}$ of Dirichlet processes, we take

$$N \sim \mathcal{DMP}(a, c, F_0) \quad \text{or} \quad G \sim \mathcal{DP}(a, F_0) \text{ with } N \mid G \sim \mathcal{MP}(c, G)$$

and

$$F_i \mid N \sim \mathcal{DP}\left(a + c, \frac{aF_0 + N}{a + c}\right)$$

for $i = 1, \ldots, n$. Borrowing ideas from Proposition 3.5, it can be proved that $F_i \sim \mathcal{DP}(a, F_0)$ marginally. Using the fact that the finite-dimensional distributions of the Dirichlet and multinomial processes are Dirichlet and multinomial distributions, respectively, the correlations induced by this

8.2 Dirichlet Process Models

exchangeable construction are the same as in Example 8.1 but with a new parameterisation, that is, for any set $B \subset \mathbb{R}$,

$$\text{Corr}\{F_i(B), F_j(B)\} = \frac{c}{a+c}. \quad (8.5)$$

and for B_r, B_s elements of the partition with $r \neq s$,

$$\text{Corr}\{F_i(B_r), F_j(B_s)\} = -\left(\frac{c}{a+c}\right)\sqrt{\frac{F_0(B_r)F_0(B_s)}{\{1-F_0(B_r)\}\{1-F_0(B_s)\}}}. \quad (8.6)$$

If we want to construct a Markov process $\{F_i\}$ with Dirichlet process marginal distribution, we take

$$F_1 \sim \mathcal{DP}(a, F_0), \quad N_1 \mid F_1 \sim \mathcal{MP}(c_1, F_1)$$

and

$$F_{i+1} \mid N_i \sim \mathcal{DP}\left(a + c_i, \frac{aF_0 + N_i}{a + c_i}\right)$$

for $i = 1, \ldots, n-1$. Borrowing ideas from Proposition 4.1, it can also be proved that $F_i \sim \mathcal{DP}(a, F_0)$ marginally. The one-lagged correlations induced by this Markov construction are the same as (8.5) and (8.6) but with c replaced by c_i.

Moreover, if we want to construct a temporal or spatial sequence of measures $\{F_i\}$, whose invariant distribution is the Dirichlet process, we take

$$G \sim \mathcal{DP}(a, F_0), \quad N_i \mid G \sim \mathcal{MP}(c_i, G)$$

and

$$F_i \mid \mathbf{N}_i^* \sim \mathcal{DP}\left(a + \sum_{j \in \partial_i} c_j, \frac{aF_0 + \sum_{j \in \partial_i} N_j}{a + \sum_{j \in \partial_i} c_j}\right)$$

for appropriate choice of subsets $\mathbf{N}_i^*$ and index sets ∂_i for $i = 1, \ldots, n$. Borrowing ideas from Proposition 5.4, Nieto-Barajas (2021) proved that $F_i \sim \mathcal{DP}(a, F_0)$. The correlations induced by this construction, for any set $B \subset \mathbb{R}$, between $F_i(B)$ and $F_k(B)$ are those given by Proposition 5.5 and coincide with (8.3), and for any two elements of the partition, say B_r and B_s for $r \neq s$, the correlation is

$$\text{Corr}\{F_i(B_r), F_k(B_s)\} = -\sqrt{\frac{F_0(B_r)F_0(B_s)}{\{1-F_0(B_r)\}\{1-F_0(B_s)\}}}\text{Corr}\{F_i(B), F_k(B)\},$$

where again $\text{Corr}\{F_i(B), F_k(B)\}$ is given by (8.3).

In all cases, the strength of the dependence is controlled by the scalar parameter c in the exchangeable case and the parameter set $\{c_i\}$ in the Markov and general dependence cases.

The general dependence sequence of random measures with Dirichlet process as invariant distribution was studied in Nieto-Barajas (2021) for carrying out Bayesian nonparametric inference. Prior correlations among measures and posterior distributions were also studied.

Appendix

Data Sets

ID	Country	Cases	Deaths	Population
1	USA	109522503	1183470	334805269
2	India	45001807	533298	1406631776
3	France	40138560	167642	65584518
4	Germany	38641986	178162	83883596
5	Brazil	38048773	707470	215353593
6	S. Korea	34571873	35934	51329899
7	Japan	33803572	74694	125584838
8	Italy	26363670	193144	60262770
9	UK	24812582	232112	68497907
10	Russia	23335550	400578	145805947
11	Turkey	17232066	102174	85561976
12	Spain	13914811	121760	46719142
13	Australia	11694319	22974	26068792
14	Vietnam	11624114	43206	98953541
15	Taiwan	10241523	19005	23888595
16	Argentina	10080046	130685	46010234
17	Netherlands	8620051	22992	17211447
18	Mexico	7649199	334472	131562772
19	Iran	7624112	146688	86022837
20	Indonesia	6814111	161920	279134505
21	Poland	6566253	119819	37739785
22	Colombia	6384891	143079	51512762
23	Greece	6101379	37089	10316637
24	Austria	6081287	22542	9066710
25	Portugal	5631281	27686	10140570

Table 1 *COVID-19 data set. Top 25 countries with more cases reported up to 29 November, 2023. Source:* `www.worldometers.info/coronavirus/`.

Pair	Time	Cens	Treat	Time	Cens	Treat
1	1	1	control	10	1	6-MP
2	22	1	control	7	1	6-MP
3	3	1	control	32	0	6-MP
4	12	1	control	23	1	6-MP
5	8	1	control	22	1	6-MP
6	17	1	control	6	1	6-MP
7	2	1	control	16	1	6-MP
8	11	1	control	34	0	6-MP
9	8	1	control	32	0	6-MP
10	12	1	control	25	0	6-MP
11	2	1	control	11	0	6-MP
12	5	1	control	20	0	6-MP
13	4	1	control	19	0	6-MP
14	15	1	control	6	1	6-MP
15	8	1	control	17	0	6-MP
16	23	1	control	35	0	6-MP
17	5	1	control	6	1	6-MP
18	11	1	control	13	1	6-MP
19	4	1	control	9	0	6-MP
20	1	1	control	6	0	6-MP
21	8	1	control	10	0	6-MP

Table 2 *6-MP data set. Remission times for children with leukemia. Freireich et al. (1963).*

Appendix Data Sets

Date	Rate	Date	Rate	Date	Rate
2006/01	0.032019	2008/12	0.045412	2011/11	0.050742
2006/02	0.035219	2009/01	0.046691	2011/12	0.050277
2006/03	0.033361	2009/02	0.050001	2012/01	0.046525
2006/04	0.033823	2009/03	0.049471	2012/02	0.053313
2006/05	0.031424	2009/04	0.052003	2012/03	0.051001
2006/06	0.035064	2009/05	0.055835	2012/04	0.048779
2006/07	0.036652	2009/06	0.052886	2012/05	0.048724
2006/08	0.036169	2009/07	0.054629	2012/06	0.048509
2006/09	0.037283	2009/08	0.057051	2012/07	0.047540
2006/10	0.036781	2009/09	0.059969	2012/08	0.049532
2006/11	0.037428	2009/10	0.054987	2012/09	0.047000
2006/12	0.037907	2009/11	0.052981	2012/10	0.048964
2007/01	0.037293	2009/12	0.052997	2012/11	0.052166
2007/02	0.038165	2010/01	0.054104	2012/12	0.049341
2007/03	0.038001	2010/02	0.051099	2013/01	0.051443
2007/04	0.036882	2010/03	0.051278	2013/02	0.048721
2007/05	0.034615	2010/04	0.054607	2013/03	0.050443
2007/06	0.035382	2010/05	0.052361	2013/04	0.051099
2007/07	0.034871	2010/06	0.051400	2013/05	0.049665
2007/08	0.035760	2010/07	0.052969	2013/06	0.050816
2007/09	0.035047	2010/08	0.050119	2013/07	0.049188
2007/10	0.036324	2010/09	0.052976	2013/08	0.048376
2007/11	0.035573	2010/10	0.053400	2013/09	0.050043
2007/12	0.035264	2010/11	0.053235	2013/10	0.049181
2008/01	0.038157	2010/12	0.055194	2013/11	0.046225
2008/02	0.036790	2011/01	0.050325	2013/12	0.047392
2008/03	0.037584	2011/02	0.052584	2014/01	0.048951
2008/04	0.036167	2011/03	0.050756	2014/02	0.047681
2008/05	0.035713	2011/04	0.052028	2014/03	0.052887
2008/06	0.036127	2011/05	0.053463	2014/04	0.049110
2008/07	0.038434	2011/06	0.055566	2014/05	0.049599
2008/08	0.037753	2011/07	0.052176	2014/06	0.048353
2008/09	0.038581	2011/08	0.052842	2014/07	0.052295
2008/10	0.040259	2011/09	0.050941	2014/08	0.048502
2008/11	0.045432	2011/10	0.048828	2014/09	0.047955

Table 3 *Unemployment data set. Monthly unemployment rates from January 2006 to September 2014. Source:* www.inegi.org.mx/temas/empleo/.

Date	Rate	Date	Rate	Date	Rate
2014/10	0.046573	2017/09	0.033394	2020/08	0.048999
2014/11	0.046656	2017/10	0.034428	2020/09	0.047893
2014/12	0.041425	2017/11	0.035032	2020/10	0.045830
2015/01	0.043784	2017/12	0.033527	2020/11	0.045309
2015/02	0.044404	2018/01	0.032882	2020/12	0.040459
2015/03	0.042562	2018/02	0.032872	2021/01	0.045507
2015/04	0.043376	2018/03	0.032737	2021/02	0.045350
2015/05	0.044256	2018/04	0.033814	2021/03	0.044629
2015/06	0.044166	2018/05	0.032410	2021/04	0.046687
2015/07	0.044941	2018/06	0.033667	2021/05	0.040527
2015/08	0.043886	2018/07	0.032951	2021/06	0.040280
2015/09	0.041950	2018/08	0.032697	2021/07	0.040646
2015/10	0.044367	2018/09	0.033318	2021/08	0.039984
2015/11	0.040678	2018/10	0.032033	2021/09	0.039128
2015/12	0.042952	2018/11	0.033467	2021/10	0.038274
2016/01	0.041601	2018/12	0.035692	2021/11	0.038372
2016/02	0.042690	2019/01	0.034659	2021/12	0.037683
2016/03	0.041353	2019/02	0.034099	2022/01	0.035867
2016/04	0.038107	2019/03	0.036305	2022/02	0.038535
2016/05	0.039928	2019/04	0.034882	2022/03	0.034490
2016/06	0.039203	2019/05	0.035495	2022/04	0.030552
2016/07	0.038098	2019/06	0.035531	2022/05	0.033031
2016/08	0.037616	2019/07	0.035258	2022/06	0.033700
2016/09	0.038429	2019/08	0.034822	2022/07	0.031659
2016/10	0.035807	2019/09	0.035183	2022/08	0.032476
2016/11	0.035943	2019/10	0.036173	2022/09	0.031225
2016/12	0.036253	2019/11	0.035388	2022/10	0.031837
2017/01	0.035311	2019/12	0.030884	2022/11	0.029848
2017/02	0.034537	2020/01	0.036624	2022/12	0.029757
2017/03	0.035406	2020/02	0.036531	2023/01	0.029093
2017/04	0.034674	2020/03	0.033113	2023/02	0.027989
2017/05	0.035437	2020/04	0.046805	2023/03	0.027927
2017/06	0.032810	2020/05	0.042622	2023/04	0.028493
2017/07	0.032551	2020/06	0.054871	2023/05	0.029505
2017/08	0.033313	2020/07	0.050024	2023/06	0.026718
				2023/07	0.028798

Table 4 *Unemployment data set. Monthly unemployment rates from October 2014 to July 2023. Source:* `www.inegi.org.mx/temas/empleo/`.

Date	IMECA	Date	IMECA	Date	IMECA
2007/01	128	2009/10	135	2012/07	115
2007/02	122	2009/11	114	2012/08	99
2007/03	144	2009/12	110	2012/09	114
2007/04	155	2010/01	126	2012/10	104
2007/05	149	2010/02	124	2012/11	156
2007/06	142	2010/03	140	2012/12	118
2007/07	135	2010/04	142	2013/01	105
2007/08	103	2010/05	151	2013/02	110
2007/09	122	2010/06	155	2013/03	115
2007/10	117	2010/07	114	2013/04	159
2007/11	118	2010/08	121	2013/05	158
2007/12	139	2010/09	101	2013/06	126
2008/01	140	2010/10	135	2013/07	109
2008/02	122	2010/11	135	2013/08	130
2008/03	128	2010/12	117	2013/09	109
2008/04	145	2011/01	96	2013/10	126
2008/05	167	2011/02	115	2013/11	101
2008/06	120	2011/03	144	2013/12	99
2008/07	106	2011/04	136	2014/01	104
2008/08	139	2011/05	151	2014/02	142
2008/09	129	2011/06	136	2014/03	146
2008/10	102	2011/07	102	2014/04	142
2008/11	123	2011/08	109	2014/05	116
2008/12	131	2011/09	113	2014/06	124
2009/01	135	2011/10	118	2014/07	96
2009/02	168	2011/11	130	2014/08	102
2009/03	156	2011/12	129	2014/09	114
2009/04	127	2012/01	108	2014/10	107
2009/05	143	2012/02	106	2014/11	120
2009/06	141	2012/03	150	2014/12	122
2009/07	121	2012/04	120	2015/01	122
2009/08	103	2012/05	139	2015/02	137
2009/09	128	2012/06	116		

Table 5 *Ozone data set. Monthly ozone measurements (IMECA) in the centered area of Mexico City. Source:* www.aire.df.gob.mx/.

ID	State	Rate	ID	State	Rate
1	Aguascalientes	0.03000	17	Morelos	0.02095
2	Baja California	0.02233	18	Nayarit	0.02034
3	Baja California Sur	0.02861	19	Nuevo Leon	0.03043
4	Campeche	0.01819	20	Oaxaca	0.01835
5	Coahuila	0.03759	21	Puebla	0.02429
6	Colima	0.02439	22	Queretaro	0.02816
7	Chiapas	0.02405	23	Quintana Roo	0.02697
8	Chihuahua	0.02709	24	San Luis Potosi	0.02647
9	Distrito Federal	0.03901	25	Sinaloa	0.02658
10	Durango	0.02984	26	Sonora	0.02555
11	Guanajuato	0.02926	27	Tabasco	0.03639
12	Guerrero	0.01150	28	Tamaulipas	0.03080
13	Hidalgo	0.02613	29	Tlaxcala	0.03468
14	Jalisco	0.02694	30	Veracruz	0.01938
15	Mexico	0.03899	31	Yucatan	0.01989
16	Michoacan	0.01855	32	Zacatecas	0.02343

Table 6 Unemployment data set. Third quarter of 2023 unemployment rates for the 32 states of Mexico. Source: www.inegi.org.mx/temas/empleo/.

State ID	Neighbours' IDs						
1	14	32					
2	3	26					
3	2						
4	23	27	31				
5	8	10	19	32			
6	14	16					
7	20	27	30				
8	5	10	25	26			
9	15	17					
10	5	8	18	25	32		
11	14	16	22	24	32		
12	15	16	17	20	21		
13	15	21	22	24	29	30	
14	1	6	11	16	18	24	32
15	9	12	13	16	17	21	22	29
16	6	11	12	14	15	22	
17	9	12	15	21			
18	10	14	25	32			
19	5	24	28	32			
20	7	12	21	30			
21	12	13	15	17	20	29	30
22	11	13	15	16	24		
23	4	31					
24	11	13	14	19	22	28	30	32
25	8	10	18	26			
26	2	8	25				
27	4	7	30				
28	19	24	30				
29	13	15	21				
30	7	13	20	21	24	27	28
31	4	23					
32	1	5	10	11	14	18	19	24

Table 7 *Neighbouring structure of the 32 states in Mexico.*

References

Banerjee, S., Carlin, B. P., and Gelfan, A. E. 2010. *Hierarchical modeling and analysis for spatial data*. 2nd ed. Boca Raton: Chapman and Hall.
Bekele, B. N., Nieto-Barajas, L. E., and Munsell, M. F. 2012. Analysis of partially incomplete tables of breast cancer characteristics with an ordinal variable. *Journal of Statistical Theory and Practice*, **6**, 725–744.
Bernardo, J. M., and Smith, A. M. F. 2000. *Bayesian theory*. 2nd ed. Hoboken: Wiley.
Besag, J. E. 1974. Spatial interaction and the statistical analysis of lattice systems (with discussion). *Journal of the Royal Statistical Society, Series B*, **36**(2), 192–236.
Blackwell, D., and MacQueen, J. B. 1973. Ferguson distributions via Pólya urn schemes. *Annals of Statistics*, **1**(2), 353–355.
Box, G. E. P., Jenkins, G. M., Reinsel, G. C., and Ljung, G. M. 2015. *Times series analysis: Forecasting and control*. 5th ed. Hoboken: Wiley.
Broemeling, L. D. 2019. *Bayesian analysis of time series*. 1st ed. London: Chapman and Hall, CRC.
Brook, D. 1964. On the distinction between the conditional probability and the joint probability approaches in the specification of nearest-neighbour systems. *Biometrika*, **51**(3/4), 481–483.
Chatfield, C. 2003. *The analysis of time series: An introduction*. 6th ed. London: Chapman and Hall.
Chatfield, C., and Xing, H. 2019. *The analysis of time series: An introduction with R*. 7th ed. London: Chapman and Hall, CRC.
Collet, J. F. 2018. *Discrete stochastic processes and applications*. 1st ed. New York: Springer.
Congdon, P. D. 2019. *Bayesian hierarchical models: With applications using R*. 2nd ed. London: Chapman and Hall, CRC.
Conover, W. J. 1999. *Practical nonparametric statistics*. 3rd ed. Hoboken: Wiley.
Cressie, N. A. C. 1993. *Statistics for spatial data*. Hoboken: Wiley.
de Finetti, B. 1937. Foresight: Its logical laws, its subjective sources. *Annales de l'Institut Henri Poincaré*, **7**, 1–68.
Diaconis, P., and Ylvisaker, D. 1979. Conjugate priors for exponential families. *Annals of Statistics*, **7**(2), 269–281.
Diggle, P. J. 2023. *Statistical analysis of spatial and spatio-temporal point patterns*. 3rd ed. London: Chapman and Hall, CRC.
Ferguson, T. 1973. A Bayesian analysis of some nonparametric problems. *Annals of Statistics*, **1**(2), 209–230.

References

Freireich, E. J., Gehan, E., Frei, E., et al. 1963. The effect of 6-mercaptopurine on the duration of steroid-induced remission in acute leukemia: A model for evaluation of other potentially useful therapy. *Blood*, **21**(6), 699–716.

Gallager, R. G. 2012. *Discrete stochastic processes*. 6th ed. Amsterdam: Kluwer Academic Publishers.

Gelman, A., and Hill, J. 2006. *Data analysis using regression and multilevel hierarchical models*. 1st ed. Cambridge: Cambridge University Press.

Gelman, A., Carlin, J. B., Stern, H. S., et al. 2013. *Bayesian data analysis*. 3rd ed. New York: Chapman and Hall.

Haining, R. P., and Li, G. 2021. *Modelling spatial and spatial-temporal data: A Bayesian approach*. 1st ed. London: Chapman and Hall, CRC.

Halmos, P. R., and Savage, L. J. 1949. Application of the Radon–Nikodym theorem to the theory of sufficient statistics. *Annals of Mathematical Statistics*, **20**(2), 225–241.

Hjort, N., Holmes, C., Müller, P., and Walker, S. (Eds.). 2010. *Bayesian nonparametrics* (Cambridge Series in Statistical and Probabilistic Mathematics). Cambridge: Cambridge University Press.

Jara, A., Nieto-Barajas, L. E., and Quintana, F. 2013. A time series model for responses on the unit interval. *Bayesian Analysis*, **8**(3), 723–740.

Joe, H. 2023. *Dependence modeling with copulas*. 1st ed. London: Chapman and Hall, CRC.

Kaplan, E. L., and Meier, P. 1958. Nonparametric estimation from incomplete observations. *Journal of the American Statistical Association*, **53**(282), 457–481.

McLeod, A. I. 1994. Diagnostic checking of periodic autoregression models with application. *Journal of Time Series Analysis*, **15**(2), 221–233.

Mena, R., and Nieto-Barajas, L. E. 2010. Exchangeable claim sizes in a compound Poisson-type process. *Applied Stochastic Models in Business and Industry*, **26**(6), 737–757.

Mendoza, M., and Nieto-Barajas, L. E. 2006. Bayesian solvency analysis with autocorrelated observations. *Applied Stochastic Models in Business and Industry*, **22**(2), 169–180.

Mood, A. M., Graybill, F. A., and Boes, D. C. 1974. *Introduction to the theory of statistics*. 3rd ed. New York: McGraw Hill.

Morones-Ishikawa, J. A., Pliego San Martin, J., García-Bueno, J. A., and Nieto-Barajas, L. E. 2021. *BGPhazard: Markov beta and gamma processes for modeling hazard rates*. 3rd ed. CRAN: R package. https://cran.r-project.org/web//packages/BGPhazard/BGPhazard.pdf.

Morris, C. N. 1982. Natural exponential families with quadratic variance functions. *Annals of Statistics*, **10**(1), 65–80.

Morris, C. N. 1983. Natural exponential families with quadratic variance functions: Statistical theory. *Annals of Statistics*, **11**(2), 515–529.

Nabeya, S. 2001. Unit root seasonal autoregressive models with a polynomial trend of higher degree. *Econometric Theory*, **17**(2), 357–385.

Nelsen, R. B. 2006. *An introduction to copulas*. 2nd ed. New York: Springer.

Nieto-Barajas, L. E. 2008. A Markov gamma random field for modeling disease mapping data. *Statistical Modelling*, **8**(1), 97–114.

Nieto-Barajas, L. E. 2020. Bayesian regression with spatiotemporal varying coefficients. *Biometrical Journal*, **62**(5), 1245–1263.

Nieto-Barajas, L. E. 2021. A class of dependent Dirichlet processes via latent multinomial processes. *Statistics*, **55**(5), 1169–1179.

Nieto-Barajas, L. E. 2022a. Bayesian nonparametric dynamic hazard rates in evolutionary life tables. *Lifetime Data Analysis*, **28**(2), 219–334.

Nieto-Barajas, L. E. 2022b. Dependence on a collection of Poisson random variables. *Statistical Methods and Applications*, **31**, 21–39.

Nieto-Barajas, L. E. & Bandyopadhyay, D. 2013. A zero-inflated spatial gamma process model with applications to disease mapping. *Journal of Agricultural, Biological and Environmental Statistics*, **18**(2), 137–158.

Nieto-Barajas, L. E., and Gutiérrez-Peña, E. 2022. General dependence structures for some models based on exponential families with quadratic variance functions. *TEST*, **31**, 699–716.

Nieto-Barajas, L. E., and Hoyos-Argüelles, R. 2024. Generalised Bayesian sample copula of order m. *Computational Statistics*, **39**, 2065–2082.

Nieto-Barajas, L. E., and Huerta, J. G. 2017. Spatio-temporal Pareto modelling of heavy-tail data. *Spatial Statistics*, **20**, 92–109.

Nieto-Barajas, L., and Núñez-Antonio, G. 2021. Projected Pólya tree. *Journal of Computational and Graphical Statistics*, **30**(4), 1197–1208.

Nieto-Barajas, L. E., and Quintana, F. A. 2016. A Bayesian nonparametric dynamic AR model for multiple time series analysis. *Journal of Time Series Analysis*, **37**(5), 675–689.

Nieto-Barajas, L. E., and Walker, S. G. 2002. Markov beta and gamma processes for modelling hazard rates. *Scandinavian Journal of Statistics*, **29**(3), 413–424.

Pan, R., Nieto-Barajas, L. E., and Craiu, R. 2024. Multivariate temporal dependence via mixtures of rotated copulas.

Plummer, M. 2023. *rjags: Bayesian graphical models using MCMC*. 4th ed. CRAN: R package. https://cran.r-project.org/web/packages/rjags/rjags.pdf.

Raiffa, H., and Schlaifer, R. 1961. *Applied statistical decision theory*. 1st ed. Cambridge, MA: Harvard University.

Regazzini, E., Lijoi, A., and Prünster, I. 2003. Distributional results for means of random measures with independent increments. *Annals of Statistics*, **31**(2), 560–585.

Ross, S. M. 2009. *Introduction to probability models*. 10th ed. San Diego: Harcourt Academic Press.

Särkkä, S., and Svensson, L. 2023. *Bayesian filtering and smoothing*. 2nd ed. Cambridge: Cambridge University Press.

Smith, A. F. M., and Roberts, G. O. 1993. Bayesian computations via the Gibbs sampler and related Markov chain Monte Carlo methods. *Journal of the Royal Statistical Society, Series B*, **55**(1), 3–23.

Spiegelhalter, D., Best, N., Carlin, B., and van der Linde, A. 2002. Bayesian measures of model complexity and fit. *Journal of the Royal Statistical Society, Series B*, **64**(4), 583–639.

Walker, S. G., and Muliere, P. 2003. Bivariate Dirichlet process. *Statistics and Probability Letters*, **64**(1), 1–7.

Index

Conjugate families
  beta & Bernoulli, 24
  definition, 23
  gamma & gamma, 25
  gamma & Poisson, 26
  list of common families, 27
  Pareto & uniform, 24

Exchangeable model
  definition, 33
  Dirichlet, 118
  Dirichlet process, 122
  gamma, 38, 39, 42
  gamma-gamma, 36
  generalisation, 43
  invariant nonparametric, 41
  invariant parametric, 36
  multivariate normal, 120
  normal, 34

General dependent model
  beyond Markov, 70
  construction via latents, 75
  exponential family, 77
  exponential family with quadratic variance, 80
  normal, 70, 72
  Poisson, 73

Hierarchical model
  definition, 44
  example, 44

Markov model
  beta, 54, 89
  binomial, 55
  Dirichlet, 118
  Dirichlet process, 122
  gamma, 56, 98
  latent variables construction, 52
  multivariate normal, 120
  normal, 58
  Poisson, 57
  survival analysis, 63
  survival analysis example, 66

Moments
  conditional, 13
  iterative formulae, 14
  marginal, 12

Probability distributions
  continuous, 8
  discrete, 4

quasi-conjugate
  binomial & binomial, 14
  Poisson & binomial, 15

Spatial model
  beta, 103, 107
  Dirichlet, 118
  Dirichlet process, 122
  latent areas, 102
  latent edges, 106
  multivariate normal, 120

Spatio-temporal model
  beta, 115, 116
  latent areas, 113, 114
  latent edges, 115

Statistical inference, 2

Stochastic process
  definition, 17
  Markov, 17
  stationary, 17

Temporal model
  beta, 84, 89, 94, 101
  gamma, 98
  moving average type, 84
  periodic, 100
  seasonal, 94

Printed in the United States
by Baker & Taylor Publisher Services